The Healing Power of Ginseng

The Healing Power
of Ginseng

Joseph P. Hou

CRC Press
Taylor & Francis Group
Boca Raton London New York

CRC Press is an imprint of the
Taylor & Francis Group, an **informa** business

CRC Press
Taylor & Francis Group
6000 Broken Sound Parkway NW, Suite 300
Boca Raton, FL 33487-2742

© 2020 by Taylor & Francis Group, LLC
CRC Press is an imprint of Taylor & Francis Group, an Informa business

No claim to original U.S. Government works

Printed on acid-free paper

International Standard Book Number-13: 978-1-138-59407-4 (Paperback)
978-0-367-21145-5 (Hardback)

Library of Congress Cataloging-in-Publication Data

Names: Hou, Joseph P., 1929- author.
Title: The healing power of ginseng / Joseph P. Hou.
Description: Boca Raton : Taylor & Francis, 2019. | Includes bibliographical references and index.
Identifiers: LCCN 2018060716| ISBN 9781138594074 (pbk. : alk. paper) | ISBN 9780367211455 (hardback : alk. paper) | ISBN 9780429489112 (eISBN)
Subjects: | MESH: Panax | Plants, Medicinal | Phytotherapy | Medicine, Chinese Traditional
Classification: LCC RS165.G45 | NLM QV 766.5.P2 | DDC 615.3/23988--dc23
LC record available at https://lccn.loc.gov/2018060716

Visit the Taylor & Francis Web site at
http://www.taylorandfrancis.com

and the CRC Press Web site at
http://www.crcpress.com

Contents

SECTION I The Herbal Medicines and the Elixir of Life: Ginseng in the World

SECTION II   The Flourishing Ginseng Business

SECTION III   Growing Your Own Ginseng Plant

SECTION IV   Data and Research

SECTION V   Composition of Ginseng

SECTION VI   The Healing Power of Ginseng

SECTION VII   How to Take Ginseng and Who Should Not Take It

Foreword

My friend Dr. Joseph P. Hou, who loves Chinese herbal medicine, understands its health-benefitting power, especially the healing nature of ginseng. Sharing over 50 years of hard work and his collection of ancient and modern scientific information, he provides a compilation of this rich knowledge in this first English edition of a masterful work on ginseng that people all over the world can share. This achievement is rare and admirable.

Ginseng is a famous and magical herbal medicine. Its use began in ancient China 4,000 years ago, but its use is now universal. This plant drug is not easily understood comprehensively and correctly.

First, ginseng as a therapeutic herbal medicine and also as a health-care product began in a long history of traditional Chinese medicine. So its use for treating disease as medicine is based on syndrome differentiation and evidence-based efficiency, which are not easy for Westerners to understand. Therefore, these profound theories need to be explained in simple language.

Second, ginseng as a plant, its different species, and its different growth environments can cause differences in efficacy; therefore, global distribution of different varieties of ginseng needs to be understood in order to choose the right kind and avoid misuse.

Furthermore, with a thorough study of the chemical composition and the pharmacologic function of ginseng, people gradually reveal and recognize the basis of the mysterious effects of ginseng. As knowledge about ginseng is more scientifically and deeply understood, ginseng will be used more correctly.

Finally, as the saying goes, the drug is a double-edged sword; correct use can be lifesaving, and incorrect use can worsen disease. Ginseng is no exception, although it is a relatively safe plant medicine.

These issues are addressed in this text. The author's excellent writing skills and orderly arrangement make reading particularly smooth and easy.

We pay tribute to and thank Dr. Hou for his contribution—we now have this masterpiece on ginseng.

You-Yu Jin, M.D.
Professor of Pharmacology
Capital University of Medical Sciences
Beijing, China

Preface

The prime aim of this book is to provide new evidence from scientific studies of ginseng, thus answering the following questions:

1. Does ginseng really work?
2. Is herbal medicine such as ginseng effective and safe?
3. How does traditional Chinese medicine (TCM) work?
4. Does ginseng really boost energy, prevent fatigue, and improve working capacity?
5. Is ginseng useful for controlling diabetes, treating erectile dysfunction, and improving longevity?
6. How can I grow my own ginseng plant?
7. What are the differences between Asian ginseng and American ginseng?
8. Where can I buy ginseng, and how much does it cost?

For more than 40 centuries the Chinese, as well as other Asians, have continuously praised the exotic ginseng root for the healing power they believe it to possess. In at least 40 diseases, including nervous and gastrointestinal disorders, all forms of debility, hypertension, diabetes, and diseases of the heart, liver, and kidney found in the classical Chinese *Materia Medica* books (*Pen-Ts'ao*), ginseng may prove to be effective. Yet what ginseng could assist you to achieve or maintain in strength, vitality, rejuvenation, and longevity, sounds more attractive to the majority of people, particularly the aging. There must be medicinal qualities, otherwise ginseng would not have been called "divine herb," "immortal herb," and "herb of the spirit" in the Orient and sold at prices 10–100 times higher than silver or gold of its own weight by Chinese druggists in the old days.

To Western doctors, traditional Chinese herbal medicine is somewhat irrational, enigmatic, and at best, empirical and nonscientific. The herb drugs are nothing but primitive remedies with a lack of predictable efficacy. However, there are many open-minded Western scientists and medical doctors who have quite different opinions, and most importantly, who do not think Chinese medicine is unfounded or based on superstition.

Professors Takagi, Watanabe, and Ishii of the University of Tokyo, said recently[1]:

> Medical science has not yet studied even a small part of the consolidated experience gained by the Chinese in the course of time. Oriental [Chinese] medicine dealt primarily with natural products rather than with synthetic ones. ... Oriental medicine employs drugs of low toxicity. The drugs do not produce a quick symptomatic effect, but rather work slowly and very often efficaciously to increase natural resistance and recuperative power of the patient.

Professor Brekhman also said that ginseng is one of the most interesting traditional remedies that merits serious study.[2]

Since World War II, ginseng has become an increasingly important plant, receiving considerable recent publicity in the press all over the world. As a result, significant changes have taken place. One of the changes is that ginseng has become the subject of intensive chemical and biological research in China, Japan, Korea, Bulgaria, the Soviet Union, and more recently, parts of Western Europe and the United States. One of the most noteworthy endeavors of ginseng research is to give the benefits of its achievements to ever wider circles of the world's population. Another notable change is the increase in ginseng cultivation centers in China, Japan, the USSR, Korea, and the United States, and the building up of quite a number of modern and collective ginseng farms for industrialized production to meet the ever-increasing international demands for ginseng tonic products and, most important of all, to lower the price.

Natural products remain the primary source of supply of many clinically useful drugs of ancient heritage. For example, at the discovery of salicin in the bark of willow trees, no one could predict that one day it could be developed as the most remarkable, most versatile, and most widely prescribed drug in history—aspirin. For centuries, the natives of India chewed snakeroot (*Rauwolfia cerpentina*) for its calming effect. Soon after the isolation of reserpine from the snakeroot, it became an exceedingly important tranquilizer and an agent in the treatment and control of hypertension. With the discovery of Mexican yams, an abundant, inexpensive source of raw material for the synthesis of numerous steroid hormones became available. Ephedra, ergot, opium, digitalis, and penicillin are other examples of many natural agents that have become contemporary therapeutic remedies. The real beauty of natural products is their low incident of toxicity, which we desire.

Today, in the search for new therapeutic agents (synthetic chemicals, of course), the Western pharmaceutical industries usually examine hundreds and thousands of potential candidates of which only a few are clinically investigated. A very small number or none of these candidates ultimately reach the marketplace. Drug companies spend an average of 5–10 years and more than $12,000,000 to develop one new drug. Even so, they still *cannot* guarantee its complete safety.

Ginseng is an age-old medicine, and its value in promoting health and happiness has long been recognized in the Orient. Yet it is a rather new wonder substance to Western people. Many people in the Western countries, particularly young Americans, have discovered ginseng only recently, and many of them are still wondering what ginseng is. This is because ginseng is not popular in the United States. Also, a literature gap exists between the East and West on ginseng. Although many legendary stories and tales about ginseng now prevail in books on herbs and medicinal plants, they are written fictitiously and erroneously to make ginseng appear more attractive and curious to the reader. None of these legends is scientific or factual. The modern scientific evidence on the pharmacologic properties and the potential clinical usefulness of ginseng published in the Orient and in the Soviet Union is totally lacking in virtually all American libraries.

Traditional Chinese medicine (TCM) originated in ancient China and has evolved over thousands of years. TCM practitioners use herbal medicines and various mind and body practices, such as acupuncture and tai chi, to treat or prevent health problems. In the United States, people use TCM primarily as a complementary health approach.

TCM encompasses many different practices, including acupuncture, moxibustion (burning an herb above the skin to apply heat to acupuncture points), Chinese herbal medicine, *tui na* (Chinese therapeutic massage), dietary therapy, and tai chi and *qi gong* (practices that combine specific movements or postures, coordinated breathing, and mental focus). TCM is rooted in the ancient philosophy of Taoism and dates back more than 2,500 years. Traditional systems of medicine also exist in other East and South Asian countries, including Japan, Vietnam, and Korea. Some of these systems have been influenced by TCM and are similar to it in some ways, but each has developed distinctive features of its own.

UNDERLYING CONCEPTS

When thinking about ancient medical systems such as TCM, it is important to separate questions about traditional theories and concepts of health and wellness from questions about whether specific interventions might be helpful in the context of modern science-based medicine and health promotion practices.

The ancient beliefs on which TCM is based include the following:

- The human body is a miniature version of the larger, surrounding universe.
- Harmony between two opposing yet complementary forces, called *yin* and *yang*, supports health; disease results from an imbalance between these forces.
- Five elements—fire, earth, wood, metal, and water—symbolically represent all phenomena, including the stages of human life, and explain the functioning of the body and how it changes during disease.
- *Qi*, a vital energy that flows through the body, performs multiple functions in maintaining health.

CHINESE HERBAL MEDICINE

The Chinese *Materia Medica* (a pharmacological reference book used by TCM practitioners) describes tens and thousands of medicinal substances, primarily plants, but also some minerals and animal products. Different parts of plants, such as the leaves, roots, stems, flowers, and seeds, are used. In TCM, herbs are often combined in formulas and given as teas, capsules, liquid extracts, granules, or powders.

TRADITIONAL CHINESE MEDICINE

In both its philosophical content as well as its practical application to the healing of disease, TCM has a great deal to offer the Western mind. First and foremost, Chinese medicine operates from the perception that illness is created as a consequence of a disturbance occurring within a person's emotional and mental bodies. Western medicine still neglects this truth. Second, TCM philosophically maintains that *healing is a process* that *must engage the entire body*—that is, regardless of where in the physical body an illness has developed, the understanding is that the entire body is ill.

For nearly 40 centuries ginseng has been universally used in TCM as a most respected and superb health-maintaining tonic remedy. It has been prescribed repeatedly by doctors for some 40 different types of ailments in China as well as other Asian countries. The miraculous power of ginseng no doubt arises from its many therapeutically effective properties.

The conventional Western view of the human body emphasizes the physical structure and components that interact in a very subtle and complex manner. The TCM model is very different. The human body is first and foremost an energy system in which various substances interact with each other to create the physical organism. The basic substances, which range from the material to immaterial, are, namely, *Yin, Yang, Qi, Jing, Shen, Blood, Body Fluid*, and *Meridians*.

In TCM and Daoist philosophy, ginseng is considered one of the top *Qi* tonics on the planet, meaning that it works to restore, cultivate, and supply *Qi* energy in the body, mind, and spirit. *Qi* loosely translates to life-force energy, which is a form of subtle energy on which the body and energy bodies run. *Qi* can be depleted in a number of ways: stress, poor diet and lifestyle, lack of deep breathing, exercise, overexertion, sexual overindulgence, and exposure to chemicals and toxins.

According to Chinese medicine and Daoist philosophy, it is this lack of— or imbalance of—*Qi* that underlies most disease and dysfunction in the body. Consequently, abundant, balanced, and flowing *Qi* energy, which ginseng helps to regulate, is believed to be at the root cause of good health, longevity, and even superhuman abilities.

Ginseng also helps to cultivate *Jing* and *Shen* energy within the self as well. *Jing* loosely translated to primordial essence, and you can think of it like a deeper, systemic version of *Qi* that is not as easily restored. "*Jing* is the deepest and most fundamental aspect of our energy. It governs the gradual processes of development and aging. *Qi* is more of a day-to-day energy, which controls our everyday functioning and overall health, and *Shen* relates to our mental, emotional, and spiritual selves.[3]" When a plant or medicine supplies *Shen* energy, it tends to have a balancing, harmonizing, and uplifting effect on our emotions, spiritual connection, and thoughts, among many other things. While all types of ginseng have effects of these three energies, each variety has subtly different properties.

In the United States, more and more health-conscious consumers, physicians, health-care professionals, and medical authorities, including the World Health Organization (WHO) and the National Institutes of Health (NIH), endorse the benefits of TCM, herbs, and acupuncture as alternative medicine. Numerous well-known physicians such as David Eisenberg, Steven Bratman, Andrew Weil, and Jonathan Wright strongly advocate the benefits of alternative medicine, and it has become increasingly important to a large segment of the American population. The author believes that TCM, herbs, and acupuncture will soon be integrated with conventional (Western) medicine in the health-care field.

Weil recently wrote:

As a practicing physician with a degree in botany for the past 12 years, I have practiced natural and preventive medicine in Tucson, AZ, using botanical remedies as one modality of treatment. I estimate that, for every prescription I write for a pharmaceutical drug, I give out 30 recommendations for botanicals. I get good results with these remedies and have seen no significant adverse reactions to them. Patients who consult me come from all over the country. They tend to be intelligent and well-educated, and many say they have given up on conventional medicine after it failed to help them attain better health. Acceptance of botanical remedies is high in this group.

At this point, some people might well ask, "Why is herbal medicine necessary? Why do we have to use ginseng? There are answers. The proper use of phytochemicals or herbal medicine for prevention or treatment of any ailments, instead of drugs, would minimize or avoid toxic side effects.

For example, ginseng could be used for flu prevention, and ginseng could improve the survival of human lung epithelial cells infected with the influenza virus.

Ginseng may help with stimulating physical and mental activity among people who are weak and often feel tired. We know for sure that ginseng is a powerful tonic medicine. A Mayo Clinic study revealed that ginseng showed good results in helping cancer patients with fatigue.

In Germany, plant drugs have a long tradition of use and are highly valued by the public. The regulatory agency, the *Bundesgesundheitsamt*, recognizes this and has made a special effort to evaluate them, not as new chemical entities, but as classic plant drugs that have a long history of use. A special expert committee, the Commission E, has been at work since 1978 and has now studied and published its findings on approximately 300 herbs. Some 200 of these have been found to have a favorable risk-to-benefit ratio and are currently approved for sale as drugs. Commission E was proactive in its determination of safety and efficacy. Its standards were those of absolute certainty of safety and reasonable proof of efficacy. The judgements it has rendered are basically sound, and the Commission E Monographs published in the *Bundesanzeiger* comprise the best formation currently available on the therapeutic use of herbs and phytomedicines.

Americans have discovered ginseng only recently, and many of them are still wondering what ginseng is. This is because ginseng is not popular in the United States. Also, a literature gap exists between the East and the West on ginseng. Although many legendary stories and tales about ginseng now prevail in books on herbs and medicinal plants, they are written fictitiously and erroneously to make ginseng appear more attractive and curious to the reader. None of these legends is scientific or factual. Modern scientific evidence on the pharmacological properties and the potential clinical usefulness of ginseng published in the Orient and in the Soviet Union is totally lacking in virtually all American libraries.

As a China-born American pharmaceutical chemist and acupuncture physician, I have been studying ginseng for more than 50 years, and I am also a ginseng enthusiast and user. From my own experience and from that of several members in my family, ginseng does indeed work. It has kept me disease free, young looking, and most important of all, able to utilize my time more efficiently and more productively to accomplish whatever I want to do. The relatively healthy and long lives of many members of my family confirm the longevity effect of ginseng. From my past 50 years of experience, it has been proven to me that ginseng is valuable in my life, and I wish its benefits to be extended to my readers as well. This book is written for this purpose.

This book is made up of 18 chapters separated into seven parts. China no doubt is the first country in the world to have discovered the wild mountain ginseng a thousand years ago and used it as folk medicine for illness. As we talk about ginseng as an herbal medicine, we cannot avoid discussing TCM. The philosophical and technical features of TCM are little heard by the American public. Fortunately, at the beginning of the twenty-first century interest in TCM or alternative medicine has rapidly expanded in the United States and other European countries. This is stated in Chapter 1.

Chapter 2 discusses the most precious herb *Panax ginseng* in China; presented are the history of the "manroot" and the medicinal value of ginseng as recorded in ancient Chinese herbal toxicity reports.

The history of the ginseng plant discovered in North America is a slightly different story than that of *Panax ginseng*. American ginseng is not as popular and has been less researched in the last hundred years. American ginseng is not popular with the American public but has been welcomed by people of the Orient, particularly the Chinese. This is discussed in Chapter 3.

Korean ginseng is botanically identical to the ginseng in China. As a result of the historically close association between Korea and China, Chinese medicine is prevalent in Korea and Korean ginseng was known to the Chinese as early as in the third century. The climate and geographical conditions may provide optimum conditions for ginseng cultivation. Korea deserves the title of the "ginseng country." This is covered in Chapter 4.

The history of Japan importing the ginseng plant from China and Korea is recorded as early as in the seventh century. Japan also then imported ginseng seeds and cultivated ginseng from Korean ginseng farms.

Ginseng is also found naturally in the mountain area of Siberia. Because its branches are spiked with thorns, it has the nickname of *touch me not* and *devil's bush*. Siberian ginseng has similar medicinal effects as *Panax ginseng*. The Chinese refer to *Panax notoginseng* as "three seven root," because the plant has three petioles with seven leaflets each. The root of *Panax notoginseng* is a very effective agent in arresting hemorrhage and bleeding in wounds, including snake and tiger bites. Internally, it is prescribed in instances of hematemesis and hemorrhage. The three plants are covered in Chapter 5.

To business-minded readers Chapter 6 will be of interest. It discusses the trade of American ginseng in the Chinese market in the past, the story of ginseng hunting. The current flourishing ginseng businesses in Korea and Japan are stated in Chapter 7.

The flourishing American ginseng business and ginseng's dollar value along with ginseng dealers and the dollar value of American ginseng are presented in Chapter 8.

If you are interested in growing your own ginseng plant, you must read Chapter 9 carefully. It provides all of the information necessary to grow ginseng.

Excitement about the cultivation of American ginseng began more than 110 years ago with the Fromm brothers in Hamburg, Wisconsin. Marathon County, Wisconsin, grows the most praised and demanded American ginseng in the world. Hsu's Ginseng Enterprises employs a combination of highly qualified staff and more than 100 during harvest season. The growing of ginseng, starting from seeding, through growing, weed control, disease control, up to harvesting, is described in detail. Interested readers should carefully read Chapter 10.

As a result of the information gap concerning ginseng between the East and the United States, the majority of the population in the United States knows little about the latest developments on ginseng. The latest developments are provided in Chapter 11.

The majority of research has been done on ginseng as well as on Siberian ginseng. Clinical and chemical research on American ginseng is yet to be done. One reason for this is the American medical establishment's skepticism of herbal remedies. American ginseng may be the best remedy for Americans whose fast-paced and energetic lifestyles would benefit from an herb that has calming and balancing effects. Details on this are provided in Chapter 12.

Chapter 13 focuses on the chemistry of all known ginsengs in the world—the active constituents contained in *Panax ginseng*, American ginseng, Japanese ginseng, Siberian ginseng, and *Panax notoginseng*. The search for economical sources of ginseng saponins from nature and even chemical synthesis may likely become the active ginseng research of the future. Controlled medical studies and research undoubtedly need to develop these natural panacea into useful, efficacious modern remedies.

Chapter 14 specifically illustrates how differences in the active chemical properties of ginseng affect medical outcomes. It is nice to see ginseng, which has been touted for centuries as a trusted herbal remedy, getting the recognition it deserves. Ginseng is called an "adaptogen," which is a substance that is supposed to help the body better cope with mental and physical stress.

Scientific research and evidence of how ginseng really works and its value as a universal tonic that promotes and maintains good health, particularly for those who are physically and mentally troubled and aged men and women, have long been recognized in the Orient and are slowly gaining recognition in the Western world. The effects of ginseng and Siberian ginseng on health and well-being and its antiaging effects are summarized in Chapters 15, 16, and 17. The results of biological (animal) and clinical (human) studies of ginseng by hundreds of scientists all over the world in the past 150 years are covered in these three chapters.

How to take ginseng, who should not take it, and possible interactions between ginseng and some of the chemical drugs are discussed in Chapter 17.

This compendium covers a wide range of information about ginseng: from the manroot story to ginseng hunting, from the two-man walk test to the inhibition of amnion cell aging, and from panaquilon to ginsenoside. This presentation of information moves from the myths to the truths about ginseng. This is the first book in the English language that has ever attempted to encompass the currently available material about the little known manroot.

A glossary, which contains the most important and commonly encountered scientific (chemical, pharmacologic, and medical) terms used in this book, is also provided.

It should be noted that I make no medical or therapeutic claims for any of the herbal remedies including ginseng listed in this book, although these herbal remedies have been used for thousands of years in the Orient and are reported safe. Like any medicine, these materials should be used with care and should be discussed with physicians who know herbal medicine.

Finally, I wish to express my gratitude to the library of the Squibb Institute for Medical Research; the National Library of Medicine; the Gest Oriental Library of Princeton University; the Office of Monopoly, Korea; and Pharmaton Limited, Lugano, Switzerland, for providing valuable references and information on ginseng research. I am also grateful to Mrs. M. Cardona for her valuable assistance in preparing the glossary of this book.

REFERENCES

1. Chen, K. K. and B. Mjkerji, eds., *Pharmacology of Oriental Plants*, Pergamon Press, New York, 1965, p. 1.
2. Brekhman, I. I. and I. V. Drdymo, *Lloyda*, 32: 46–51, 1969.
3. Hou, J. P. *The Myth and Truth About Ginseng*, Cranbury, NJ: AS Barnes and Co., Inc., 1978.

Author

Joseph P. Hou, Ph.D., AP, holds a doctorate in pharmaceutical sciences and chemistry from the University of Wisconsin–Madison. He is the author of four books: *Myth and Truth about Ginseng* (A.S. Barnes, 1978); *Alternative Therapies for Cancer and Common Ailments* (1st Books Library, 2002); *The Healing Power of Chinese Herbs and Medicinal Recipes* (Haworth Integrative Healing Press, 2005); and *Health Longevity Techniques* (Author House, 2010). Hou has published numerous scientific papers and holds several U.S. new drug patents. Since 1983 he has been the president of the J.P. Hou Institute for Pain and Integral Health in Orlando, Florida. Hou is the former president of the Florida Acupuncture Association and a former vice chair of the Florida Acupuncture Board. He also lectures on complementary/alternative medicine, acupuncture, and Chinese herbal medicine.

Section I

The Herbal Medicines and the Elixir of Life

Ginseng in the World

1 The Chinese Style of Healing
The Best Holistic Medicine and Herbal Medicine

TRADITIONAL CHINESE MEDICINE

Traditional Chinese medicine (TCM) or *Chung-I*, is the daily medical practice in China. *Chung-I* incorporates traditional techniques and methods as well as herbal medicine.

The medical agents used for healing are primarily natural products rather than synthetic chemicals. These natural products are derived not only from herbaceous and arborous species, but also from animal and mineral sources. The Chinese physician tailors his treatment to fit the individual patient's condition as well as the nature of the disease. His prescriptions usually contain many medicinal substances to produce synergistic action, or to neutralize negative interactions, and to mask the bitter taste of the crude drugs. These drugs are administered, in most cases, in the form of a decoction.[1,2] Chinese medicine recognizes complex mixtures of drugs and does not appreciate the advantage or effectiveness of a single drug entity. This is a fundamental difference between Chinese and Western medicine.

Acupuncture, the insertion of slender solid needles into specific points in the body, is a rather elaborate technique of *Chung-I*. Acupuncture therapy has also been used in Korea, Japan, and other Southeast Asian countries for thousands of years, and in France for more than a century. A small but growing group of British physicians use acupuncture. It is practiced in Germany, and as long as their love affair with China endured, the Russians used acupuncture and conducted extensive research on acupuncture. And finally, the people of Canada and of the United States suddenly, and with characteristic fervor, have discovered acupuncture.[3] Acupuncture analgesia, a highly sophisticated medical technique developed in China only 15 years ago, has been more impressive and convincing than therapeutic acupuncture. In China it has now been accepted and is used in surgery in nearly all general hospitals.[4]

In the last few years, the term *acupuncture* has become familiar in the United States. The question raised in 1962 about its usefulness has been answered by a great number of American doctors who have seen it applied in the People's Republic of China. Most of these Western observers have been astounded by the extraordinary effect of the acupuncture needles in producing surgical anesthesia, profound enough for the performance of surgical operations.

The New York Times correspondent James Reston experienced acupuncture therapy in a Peking general hospital. The journalist, as well as all other American physicians who subsequently visited China, observed and testified to the fact that acupuncture was "verity" and "no delusion." They brought back from the People's Republic of China irrefutable proof in the form of films depicting the performance of major surgery under acupuncture analgesia with immediate and complete postsurgical functioning of the patient.[5]

Although its application in surgery has not been as widely applied as acupuncture therapy in TCM, some special surgical techniques do exist. A method of cataract surgery called "coughing," with a site of scleral incision, had been used centuries ago in China.[4]

In addition to herbal medicine, acupuncture, and surgery, physical therapy, diet therapy, and massage are other features and techniques of traditional *Chung-I*. Massage was fully developed during the T'ang dynasty. The simpler hand movements of massage supposedly stimulate circulation and make muscles and joints supple, thus reducing pain. Diet therapy is even less known to Westerners, although it is an indispensable part of Chinese medicine. The first compendium on diet therapy, written by Sung Ssu-Miao, was published in the T'ang dynasty; several other related books on diet therapy were also published during the late T'ang and Sung dynasties. These books discuss the properties, uses, and amounts of specific food substances for diet therapy and prevention of diseases.

Practitioners of *Chung-I* (TCM) are generally divided into herbalist and acupuncturist groups. The former turn to an enormous *Materia Medica* based on the pharmacological properties of herbal remedies. *Ginseng* (*Panax ginseng*) is the most popular drug and has been prescribed in practically every prescription—for weakness, debility, fever, chills, cough, pallor, and of course, to enhance virility and restore health.

The practice of traditional medicine is followed not only in China, but also in Japan, Korea, Taiwan, Indochina, Singapore, and other parts of Southeast Asia, and by at least 18 million overseas Chinese. Thus, a third of the world's people (the Orientals) receive some form of TCM. For this reason, the term *Oriental medicine* has been used by Westerners to describe the traditional medical practice in the Orient.

In Chinese history,[6,7] the three legendary emperors Fu-His, Shen-Nung, and Huang-Ti are the founders of early Chinese civilization. To Fu-His is attributed the *Canon of changes* or *I-Ching*, regarded as the most ancient Chinese philosophy and medicine. Shen-Nung, also known as Yin-Ti, is the father of agriculture and herbal medicine. It is said that he tasted hundreds of herbs and other crude drugs in order to acquaint himself with their properties and usefulness. He is commonly attributed the compilation of the first *Pen-ts'ao*, or Chinese *Materia Medica*. Huang-Ti, the Yellow Emperor, contributed a complete treatise on the principles of health and medicine in 2697 BC, known as *Huang Ti Nei Ching Su-Wen* (*The Yellow Emperor's Classic of Internal Medicine*) or simply called *Nei Ching* (*The Canon of Medicine*), which consisted of 18 volumes with 162 chapters.[7] Although it was written more than 4,000 years ago, it has been recognized as a most valuable treatise on internal medicine and supposedly the world's oldest *extant* medical book. TCM can claim to be the world's first organized body of medical knowledge.[8]

The Canon of Medicine was also the most interesting medical book that ever existed. It was compiled in the form of a dialogue between the emperor and his physician minister, Ch'i Pai.[7] Their discussions included the philosophy of nature, theories of yin and yang, the Five-Elements doctrine, pulse diagnosis, mechanisms of viscera, the vascular system, the value of life, and achievement of the perfect body. The same book also illustrated that the prevention of disease can be achieved by regular habits, a proper diet, a suitable combination of work and rest, and the maintenance of a peaceful mind.

EARLY MEDICAL PHILOSOPHY

TCM is also the most pervasive and the most unyielding of the indigenous systems. It is based on the tenet that a human being is a microcosm constantly interacting with the immense universe, which influences and also controls every aspect of that person's life, including his or her health.[3]

As a matter of fact, early Chinese medicine incorporated philosophy and religion. Three essential religious philosophy concepts that control early medical thinking are *Tao, yin-yang theory,* and the *doctrine of Five Elements.*[7]

TAO

During the sixth century BC, Lao Tzu, the spiritual father of Taoism, founded the natural philosophy. Taoism is a concept common to all Chinese. It is the key to the mysterious intermingling of Heaven and Earth. Tao means "way" and the method of maintaining the harmony between this world and the beyond. As in an agricultural society, the ancient Chinese philosophy is always related to nature and cosmology. The only manner in which man could attain the right Tao was by emulating the course of the universe and adjusting completely to it.

Tao plays an important role as the regulator of the universe and the highest code of conduct. A man's health and longevity depend on his behavior toward Tao. Longevity itself became to a certain degree a token of sainthood, since it was an indication that it had been achieved by personal effort of complete adherence to Tao. Those who follow Tao achieve the formula of perpetual youth and maintain a youthful body.

YIN AND YANG THERAPY

Yin and yang are the shady side of a hill (yin) and the sunny side of a hill (yang) from literary translation. Yang stands for sun, heaven, day, fire, heat, dryness, light, and many other positive and masculine subjects, while yin represents moon, earth, night, water, cold, damp, dark, and many negative and feminine subjects. Yang means motion, hence, life; yin means standstill, hence, death. The principle of yin and yang is the basis of the entire universe. It is the principle of everything in creation. However, it must be borne in mind that yin and yang are conceived as one entity and that both together are ever present. Day changes into night, spring and summer change to autumn and winter, light changes into dark, and so on. From these striking manifestations, it was deduced that all happenings in nature as well as in human life were conditions caused by the constantly changing relationship of yin and yang.

Heat is yang, and cold is yin. In the human body excessive yang causes fever, and excessive yin causes chills. Every food or medicament has a predominant character, either of yin or yang. The art of healing is to ascertain where and in which direction the *equipoise* of yin or yang has been lost in the balance, then the appropriate medication or treatment has to be applied to restore it to normal, and to restore internal balance and harmony. This is the essence of Chinese medical thinking.

The *Nei Ching* provides us with many examples of this interchange between yin and yang and of the duality preserved within a single thing. As to the interrelation of yin and yang in man, male belongs to yang, female belongs to yin, yet both male and female are products of the two elements, and hence, both qualities are contained in both sexes. In the dual nature of yin and yang within the human body, yin and yang correspond to the surface and the interior, respectively. The yin and yang in harmony means health; disharmony or undue preponderance of one element brings disease and death. Man received the doctrine of Tao as a means of maintaining perfect balance and securing for himself health and long life.[7]

DOCTRINE OF FIVE ELEMENTS

More tangible components of yin and yang are the five elements. Yin and yang, in addition to exerting dual existence, are subdivided into metal, wood, water, fire, and earth—the five elements. Man was said to be the product of heaven and earth by the interaction of yin and yang, and therefore contains the five elements.

The sequence of the five elements varies according to the viewpoint from which they are enumerated. The *Nei Ching* explains the mutual victories of the five elements as follows:

Wood brought in contact with metal is felled.
Fire brought in contact with water is extinguished.
Earth brought in contact with wood is penetrated.
Metal brought in contact with fire is dissolved.
Water brought in contact with earth is halted.

The sequence of subjugations is that metal subjugates wood, water subjugates fire, wood subjugates earth,0 fire subjugates metal, and earth subjugates water.

The doctrine of Five Elements also extends to grains, fruits, animals, vegetables, flavors, odors, climates, musical notes, human organs, and many groups, each of which contains five components. The five grains that act as nourishment are wheat, glutinous millet, millet, rice, and beans. The five fruits are peaches, plums, apricots, chestnuts, and dates. The five domestic animals that contribute additional nutrients are fowl, sheep, beef, horses, and pigs. The five vegetables are mallows, coarse greens, scallions, onions, and leeks. The human body contains five viscera: liver, lungs, heart, spleen, and kidneys.

The *Nei Ching* gives the following explanations of how the natural elements affect the human body. The climate elements affect the viscera of our body: heat injures the heart, cold injures the lungs, wind injures the liver, humidity injures the spleen, and dryness injures the kidney.

The five viscera, of course, control the body: heart controls the pulse, lungs control the skin, liver controls the muscles, spleen controls the flesh, and kidney controls the bones. The five spiritual resources of our body are also controlled by the five viscera: liver controls the soul, heart controls the spirit, spleen controls ideas, lungs control the animal spirit (ghost), and kidney controls the will.

The five flavors affect the body in the following manner: salty flavor hardens the pulse, bitter flavor withers the skin, pungent flavor knots the muscles, sour flavor toughens the flesh, and sweet flavor causes aches in the bones. The five flavors are said to be effective not only on the five viscera but also on all parts of the body that are connected with the five viscera. If people pay attention to the five flavors and blend them well, their bones will remain straight, their muscles will remain tender and young, breath and blood will circulate freely, pores will be fine in texture, and consequently, breath and bones will be filled with the essence of life.[7]

FEATURES AND TECHNIQUES OF TRADITIONAL CHINESE MEDICINE

DIAGNOSIS

Diagnosis in Chinese medicine, or *zue-chen*, is listed in *Nei Ching*. There are four basic methods used to diagnose a patient. They are visual observation, questioning about case history, auditory systems, and taking the pulse. Chinese *pulsology*, developed in great detail by Wang Shu-ho, recognizes three spots along the *radial artery* of each wrist, detected with the tips of three fingers. The pulse readings reflect the functioning of different viscera. In Chinese medicine, pulse-taking has been the chief method of diagnosis. At the three pulse spots each wrist has a deep and a superficial reading, thus giving a total of 12 different pulses. If the patient is a female, the right radial artery is palpated first; if a male, the left is palpated first. The rate, strength, and direction of the beat in each segment of the pulse are determined. A strong pulse indicates a yang-type disease, while a weak pulse represents a yin-type disease. Only experienced doctors can diagnose malfunction of any part of the internal organs. Obviously this is quite a different diagnosis technique from that in Western medicine. The art of pulse diagnosis is extremely complex in the way it works. This has been the chief point of controversy between those who understand and those who do not.

ACUPUNCTURE

Acupuncture is another important branch of Chinese medicine. The therapeutic art of acupuncture (not acupuncture anesthesia) had been fully developed by the time of *Nei Ching*, in which elaborate descriptions of the practice and principles were recorded.

The rationale of acupuncture historically has been based on stimulation of a *Ching-lo* system. *Ching-lo* is a system of channels and ducts that is anatomically distinct from the circulatory or nervous systems. Acupuncture has been used to cure many diseases. It is used most spectacularly to alleviate deafness and induce anesthesia. A publication from the Research Institute in Beijing makes the following grandiose claims:

> Preliminary observations show that acupuncture…can exert influence on the visible elements of the blood, the peristalsis and secretion of the stomach and intestines, and the

secretion of bile; stimulate the kidney's power of excretion, and improve the conditions of blood pressure and cardiac impulse, increase the amount of immunizing agents in the body, and stimulate the kidney's power of excretion, and improve the conditions of blood pressure and cardiac impulse, increase the amount of immunizing agents in the body, and stimulate cytocannibalism…produce curative effects of the central nervous system, segmental reflex, blood, and local parts of the body.[3]

In the classical system of Chinese medicine, acupuncture is used in both diagnosis and treatment. The theoretical basis is the system of acupuncture points distributed along the meridian, each linked with one or more internal organs. At least 361 acupuncture points have been identified. The fundamental concept of acupuncture is that detailed knowledge of the distribution of these points allows both diagnosis and treatment. Malfunction of an organ may be recognized by hypersensitivity of the corresponding acupuncture points and relieved by stimulation at the irritable point. In general, arthritis, neuritis, and paralysis after nerve injury or stroke are among the disorders treated most successfully. In these conditions the needles are stimulated electrically, usually with a square-wave alternating pulse at about 200 cycles per minute. The power source is a 6-volt battery. This stimulation causes obvious muscular twitching at the site of insertion of the needle but apparently no great discomfort for the patient. Acupuncture therapy has also been widely used in the treatment of angina, to relieve cardiac pain, for analgesic effect in childbirth, in migraine, and in many chronic skeletal disorders.[8]

China's Xinhua News Agency reported that acupuncture anesthesia was successful in 95% of the trials on 360 horses and other animals. The fact that so many horses, mules, donkeys, cattle, and pigs have responded to the method should be an eloquent denial of some skeptics' visioning that success with human beings has been due to psychological preparation (Xinhua News Agency, November 20, 1971). Other reports from the veterinary medicine field state that acupuncture will increase milk production in cows and the speed of racing horses.[3]

EXTERNAL MEDICATION

As to external medication, certain manipulative arts *Teri Ma* (notable massage) for reduction of dislocation, treatment of fractures, and bone setting have been carried out with a high degree of practical skill. The traditional methods of treating fractures are very different from those in the West. In the traditional way, reduction of the fracture is achieved slowly, using a short splint fixed with bandages over soft paper or cloth padding. The splints maintain alignment of the fracture, but they are not designed to immobilize the joints above and below it. The rationale of the treatment is that the muscles and joints around the fracture should retain their mobility; exercise is encouraged, and the splints are adjusted as often as necessary to cope with the progressive reduction in the swelling around the injury. It is said that in simple fractures for which it is best used, traditional treatment achieves rapid relief of pain and swelling and allows early mobilization; the fractures heal more quickly than those treated in plaster, and the functional results are better.[8]

CUPPING AND MOXIBUSTION

These are less well-known techniques to Western physicians; both are age-old medical techniques in the treatment of counterirritation. Cupping is used for conditions such as backache, in which the mouth of a glass cup is heated and then applied to the selected area of the skin. Cooling of the air inside the cup forms a partial vacuum so causing suction and the formation of a hematoma. Moxibustion is another technique occasionally used in TCM. The treatment is conducted in conjunction with acupuncture. After insertion of the acupuncture needles into the afflicted area of the body, a ball of moxa is applied to the outer end of the needle and ignited. The moxa ball burns for a few minutes, emitting a fair amount of both heat and smoke. Good results have been obtained in patients with frozen shoulder, sciatica, and musculoskeletal disorders.[8]

CHINESE HERBAL REMEDIES

The Chinese consistently refer to their herbal remedies as being the product of 4,000 years of the people's struggle against disease, and cite the millennial endurance of these remedies as adequate empirical proof of their effectiveness. The raw materials of traditional herbal remedies include not only more than 1,500 plants—dry leaves, barks, fruits, roots, stems, flowers, seeds, and nuts—but also such exotic objects as snakes, newts, stalactites, agates, and antlers. These remedies are given as teas, decoctions, infusions, or as a mixture compounded from the raw materials after processing by much the same techniques as those used by herbalists in the West—that is, drying, shredding, grinding, boiling, or infusing with certain solvents to extract the active principles. More than 10 types of pharmaceutical preparations, such as powders, granules, pills, masses (in the form of a golf ball), solutions, tinctures, medicated dressings, adhesive plasters, ointments, and pastes, are commonly found in the classical Chinese drugstores. Usually, the granules, pills, and masses are coated with preservatives or stored in beeswaxed shells to preserve the drug from oxidation and moisture decomposition.

EVOLUTION OF HERBAL MEDICINE

It is believed that in ancient times, while fighting for survival, the Chinese and people in other parts of the world must have acquired experience in selecting naturally available materials and concocting healing potions to eliminate pain, reduce fever, control suffering, counteract diseases, and heal wounds. By trial and error, they gained practical knowledge that was useful in determining what minerals, or what animals, and which parts of specific plants possess the desired healing activities and which ought to be discarded because of their toxicity. Knowledge concerning their medicinal properties and instructions as to their correct uses has been handed down from one generation to the next. In the course of time, a substantial volume of information of herbal remedies was thus accumulated. These crude drugs were classified and compiled in a compendium known as herbal or *Materia Medica*, and the particular name of *Pen-ts'ao* is used in Chinese medicine.

In the Western civilization, Hippocrates (460–370 BC) has been referred to as the Father of Western Medicine.[9,10] Dioscorides wrote his *De Materia Medica* in AD 78. In it, he described 600 plants that were known to have healing power. Galen (AD 131–201) was most famous for his experience in herbal medicines and his pharmaceutical preparations. Yet it was more than 500 years ahead of Galen that the Chinese *Pen-ts'ao* had been fully developed and systematized.

Numerous valuable drugs were derived from nature not only in China but in other parts of the world as well. Morphine, digitalis, quinine, atropine, ephedrine, and reserpine are examples of outstanding remedies developed from herbal medicine.

In the People's Republic of China, most of the herbs grew wild in the past. Herbs are grown on state or national farms today. The people have been encouraged to exploit the mountainous areas that are most suitable for cultivation of herbs. Provinces in South China (Kwang-tung, Kwang-si, Yun-nan, Kwei-chow, and Sze-chwan) are the leading medicinal herb–producing regions. The experimental plantation of the Institute of Materia Medica located in the western outskirts of Peking is one of the biggest government-controlled medicinal herb cultivation centers. About 1,500 species of herbs are under cultivation in an area of about 64 acres. In past years, about 10,000 "barefoot doctors" (health aides with basic-level training) have been trained there each year to be thoroughly familiar with the commonly used herbal remedies. The experimental station maintains cultivated ginseng plants with special attention paid to sunlight exposure and diseases.

PEN-TS'AO

China is perhaps the richest country in the world in natural medicine. These medicinal agents have been studied by scholars and physicians in the past 40 decades, classified, and recorded systematically in the compendia called *Pen-ts'ao*. The *Pen-ts'ao* in Chinese medicine is actually a combination of *Materia Medica*, pharmacology, and pharmacopoeia. Although Shen-nung is commonly attributed the editing of the first *Pen-ts'ao*, it was not until the Later Han dynasty that the great work of *Shen-nung Pen-ts'ao Ching* was formally published as the first official dispensatory (pharmacopoeia) in Chinese history. It contains 365 different medicaments, 237 from botanical, 65 from animal, 43 from mineral sources, and 20 of unknown origin. According to their properties and usefulness, these medicaments were classified into superior (first), middle (second), and inferior (third) classes. The superior drugs, including ginseng, are absolutely nontoxic and are used for a wide variety of diseases; the middle-class drugs are effective for limited diseases and are slightly toxic; the inferior or the third-class drugs are useful only for particular sicknesses and should be used with caution because of their potent activities and toxicities.[11,12]

According to the definition of the upper class of drugs, putting ginseng into the category of a natural health food is entirely proper, too. It is very characteristic that ginseng has dualism as a drug and a natural food. Such dualism has been thought to be an ideal feature of a panacea in Chinese medicine.

We can describe the characteristics of Chinese drugs, such as ginseng, in terms of the prevention of ills which precedes the curing of ills. They can normalize the living body of man through the adjustment of functions of the whole body. What brings a long duration of youth and longevity is natural drugs, and ginseng is the best of all natural drugs.

During the Liang dynasty (about AD 500), a great physician, T'ao Hung-ching (AD 452–536) was born at Mo-ling (now Nanking), and he received exceptional gifts from an early age and devoted himself to the practice of medicine.[6] He re-edited the great *Shen-nung Pen-ts'ao Ching* into *Shen-nung Pen-ts'ao Ching Chichu*. He also composed another book called *Ming-i Pieh-lu*, which contains 365 new and effective medicaments and many prescriptions that had been praised and used by many eminent physicians for more than 1,000 years during the earlier Chou (1122–255 BC), Ch'in (255–209 BC), the Former Han (209 BC–AD 23), the Later Han (AD 25–220), and Wei (220–543) dynasties. Hung-ching was the first who gave the descriptions of Chinese ginsengs (from Koguryo and Paekche), their pharmacological properties, and the methods of preserving them.

At the end of the Han dynasty, a system of Chinese medicine was firmly established. A well-known physician, Chang Chung-ching, wrote his famous *Treatise on Fever*. Chung-ching was one of the three greatest physicians of the Han dynasty. The other two outstanding doctors were Hwa T'o and Tsang Kung. Chung-ching conducted extensive clinical trials with ginseng and invented many prescriptions containing ginseng properties (odor, color, taste, etc.); his *Treatise on Fever* presented indications and pharmacological properties; clinical applications, contraindications, and precautions; and prescriptions, formulations, and dosages of the drug. This great writing has been praised the world over, and beginning in the seventeenth century, it has been translated into Latin and every major Eastern and Western language.[11]

The greatest pharmacologist and physician Li Shizhen (1515–1593) seriously studied more than 800 commonly encountered medicaments and traveled thousands of miles in order to collect known and unknown medical herbs. He carefully examined past editions of *Ben Cao* written in the Tang, Song, and Southern Song dynasties, and based on his own rich knowledge and experience gained during 30 years of hard work finally composed a monumental masterpiece, *Ben Cao Gang Mu* (*Compendium of Materia Medica*). This book was published in 1596, 3 years after his death.

Also during the Ming dynasty, Western civilization was introduced into China through Jesuit priests. Father Matteo Ricci was the first to introduce Western medicine to the Chinese emperor in 1601. Later in the Ch'ing dynasty (1644–1912), most of the Western medical books and theories were introduced into China. As a result, Chinese traditional medicine then became a disaster under the pressure of the Western powers.

From the Later Han to the Ch'ing dynasty, more than 50 kinds (editions) of *pen-ts'ao* or dispensatories (including compendia on diet therapy) were published officially and nonofficially. These represent the world's richest knowledge of herbal medicine.

It is correct to say that traditional Chinese herbal medicine is not just a class of folk medicine. It is a well-organized medical system developed by individual physicians and government institutions based on cumulative experience and clinical trials.

MAO'S MODERN CHINESE MEDICINE

Since the nineteenth century, the age-old Chinese civilization, of which traditional medicine was an integral part, had started to disintegrate under the impact of Western pressure. The Western missionary movement, specifically the Jesuits at

Beijing, brought Western medicine to an isolated, feudal China. Besides the Western missionary establishment, the Western military, economic, and political powers forced China to accept Western culture, commerce, and medicine. Particularly after the Opium War (1839–1842), all of the Western powers secured a foothold in Chinese territory. The Manchurian government (the Ch'ing dynasty) soon collapsed, and a new government, the Republic of China, was born in 1912.

During the Republic of China period (1912–1949), however, the (KMT Party) government's decision was to sponsor only Western medicine, while the traditional medical practice still persisted among the majority of the people, and the fate of Chinese medicine was seriously threatened and oppressed. The May Fourth Movement of 1919 even denounced traditional medicine as nothing but noxious, backward, superstitious, irrational, and nonscientific, which had to be destroyed for China to survive in the modern (Western) world. Further, the government then published a decree officially prohibiting the practice of Chinese medicine. Soon civil war started, and the military Japanese invaded China. The whole country turned into a mass disaster.

Civil war ended in 1949 with the victory of the Communist party. The new government, the People's Republic of China, without delay set up a different policy on medical practice in China that took into account the traditional art of medical practice, and in addition, the herbal medicine and Chinese formulary was to be restudied extensively by modern scientific methods.

In 1950, Chairman Mao clearly expressed again the same medical policy that he had announced during the anti-Japanese war. In a directive to the First National Health Conference in Beijing, he stressed that the Chinese should "unite all medical workers, young and old, of the traditional and Western school trained, and organize a solid united front to strive for the development of the people's health work."[13] The essence of Mao's medical thought was that (1) medical and health work should be focused on the rural areas; (2) medical services must be first for the working class; (3) practitioners of TCM must be united with Western school–trained doctors; and (4) health work must be integrated with mass movement. The ultimate goal of Mao's drive was to produce a unified system of modern Chinese medicine and Western medicine.

Accordingly, the major effort of medical practice has been to send teams of urban doctors from large medical centers on periodic tours of duty in the countryside, thus providing more and better medical service for the relatively poor and neglected peasantry. Simultaneously, there has been a drive to train a huge number of "basic level health workers," the "barefoot doctors" for the mass population. Since the People's Republic of China was built up with the working class, to provide them with suitable medical care is essential.

It is crystal clear that Mao wanted to preserve TCM, develop it, and integrate it with modern Western medicine. On many occasions he succinctly expressed that "Chinese medicine is the summation of the experience of the Chinese people in their struggle against diseases over the past 5,000 years.... Chinese medicine is a great treasure bank which must be explored and further improved with our new efforts." He further announced that "toward world health program [that the] Chinese make a great contribution is without question, one of these is the Chinese medicine.[14,15]

In addition to the new public health drive, scientific research at all levels on herb medicine as well as new antibiotics has made equal progress in the past 20 years. From

1950 to 1960, more than 10,000 scientific and clinical papers were published just on studies of herbal medicine. Few herbal medicines have antibacterial activities; the search for and manufacture of new antibiotics is similarly urgent. The government is also trying to put production on a more scientific footing by growing important herbs such as *ginseng* on a plantation basis rather than depending on forest collection. In the past 25 years, Chinese scientists have also developed a number of herbal drugs with encouraging results for the treatment of coronary diseases, appendicitis, gallstones, heart diseases, burns, birth control, and cancer.[16]

The Institute of Materia Medica in Beijing, under the auspices of the Chinese Academy of Medical Sciences, was established in 1958 for purposes of research on new herbal drugs and new principles for common diseases. The Shanghai Institute of Materia Medica, the Institute of Organic Chemistry, and the Institute of Biochemistry in Shanghai are other leading research centers on modern Chinese medicine. The present research on herbal remedies is a national dedication to follow Chairman Mao's exhortations to explore China's herbal drugs.

Recently, E. Grey Diamond, director of the Cardiovascular Center in Kansas City, Missouri, said on his return from his second trip to China in 2 years, "I was extremely impressed on the tremendous amount of information in the herb medicine. Not a cult, the Chinese had recorded and codified their herbal medicines, whereas the West had abandoned analysis of botanical medicine in favor of synthesis." Regarding the use of herbal medicine for the heart-troubled patient, he said, "I have case histories of heart disease which herbal medicines being extremely effective." Finally, Diamond speculated that "I predict that herbal remedies would follow acupuncture as the next medical import from the People's Republic of China."[17]

PEOPLE BEGIN TO ACCEPT ALTERNATIVE MEDICINE WORLDWIDE

In recent years, the general public in the United States as well as other Western countries in Europe has started to accept the value of alternative medicine. This is because current chemical drug conventional medicine is in a state of terrible disarray.

At the beginning of the twenty-first century, interest in alternative medicine is rapidly expanding. TCM and acupuncture can deliver holistic health care. To elderly people, when conventional medicine offers no help for their pains, chronic ailments, and suffering, acupuncture or TCM can play an important role in improving their health.[18]

In the United States historically, the Office of Alternative Medicine (OAM) was established in 1992 within the organization of the National Institutes of Health (NIH). The NIH is one of eight health agencies within the U.S. Public Health Service of the U.S. Department of Health and Human Services (DHHS). NIH is the federal government's focal point for biomedical research in the United States. Based on necessity, the OAM was established to facilitate and coordinate the evaluation of alternative medical treatment modalities through research projects and other initiatives with NIH's institutes and other centers. At that time, OAM's primary role was to emphasize the vigorous scientific evaluation of alternative therapies and establish a clearinghouse to provide information to the public. The OAM was upgraded 6 years later, and the National Center for Complementary and Alternative Medicine (NCCAM) was initiated in 1998 by the U.S. Congress and signed by President Bill

Clinton on October 21, 1999. The purpose of NCCAM was stimulating, developing, and supporting research on complementary and alternative medicine (CAM) for the benefit of the public. The NCCAM is one of more than 27 institutes and centers (ICs) composing the NIH.[19] And in December 2014, the U.S. Congress, in a measure signed by President Barack Obama, changed NCCAM's name to the National Center for Complementary and Integrative Health (NCCIH, under US Public Health Services) to more accurately reflect the center's commitment to studying various health approaches.

The general public and the government health agencies accept alternative medicine means in order to accept traditional Chinese medical practice (TCM). The following briefly explains how TCM works.

THEORY OF TRADITIONAL CHINESE MEDICINE

TCM emphasizes the proper balance of bioenergy, or *Qi* (pronounced "chee"), in health and disease. TCM consists of a group of therapeutic techniques and methods, which include acupuncture, moxibustion, herbal medicine, massage, dietary therapy, and breathing (Qigong) therapy. Chinese medicine views the body as a self-repairing mechanism, and treatment enables the body to return to its natural balanced state of health.

According to TCM, there are 14 major pathways, called *meridians of bioenergy*, in the human body. Chinese medicine practitioners believe that these meridians conduct *Qi* or bioenergy between the surface of the body and internal organs. *Qi* regulates physical, mental, and emotional balance. When the flow of *Qi* is disturbed through poor health habits, poor nutrition, emotion, aging, or other disturbances, pain and/or disease can result. Acupuncture, manipulative massage, *Tai Ji* exercise, and proper diet or tonic herbs can help to keep the flow of bioenergy, *Qi*, unblocked.

Acupuncture is an important branch of TCM. Acupuncture is the practice of inserting thin needles into subcutaneous connective tissue or muscle at specific points in the body. The specific points are called *acupoints*. These points are used by a skilled acupuncture physician to treat a wide variety of illnesses. Acupuncture unarguably gives pain relief (analgesia effect) to a significant number of people. In addition, proper diet and medical herbs are essential in the practice of TCM.

Although the exact number of people who use TCM in the United States is unknown, it was estimated in 1997 that some 10,000 practitioners served more than 1 million patients each year. According to the 2007 National Health Interview Survey (NHIS), which included a comprehensive survey on the use of complementary health approaches by Americans, an estimated 3.1 million U.S. adults had used acupuncture in the previous year. The number of visits to acupuncturists tripled between 1997 and 2007. According to the 2007 NHIS, about 2.3 million Americans practiced tai chi and 600,000 practiced qi gong in the previous year.

TCM is now practiced in one form or another by more than 300,000 practitioners in over 140 countries. The first hospital for Chinese medicine in Europe was opened in Germany in 1990. British general practitioners are increasingly contracting out for acupuncture services, public health-insurance companies in Germany routinely refund part of the costs of acupuncture treatment provided by trained doctors, and in France, acupuncture is a widely accepted part of health-care provision. Degree programs in

Chinese medicine are now offered at several British universities, and courses in TCM are established at European medical schools. Tong ren Tang, Beijing's oldest pharmacy founded in the seventeenth century, opened a branch in central London in 1995.

Given the worldwide hegemony of biomedicine throughout most of this century, the global emergence of TCM is a monumental event. Proponents already speak of the twenty-first century as the "century of TCM." More likely it is a continued integration of TCM into existing health-care systems worldwide.

A remarkable success of alternative medicine is the success of herbal medicine for treating malaria patients. On October 5, 2015, Chinese herbal researcher Tu YouYou won the Nobel Prize in Medical Sciences for discovery of the antimalarial herbal drug Artemisia[21], which has saved millions and millions of malaria patients living in South Asia and Africa. It was reported that she worked on the crude herb *Artemesia apiacea* using different solvents at different temperatures, and although she failed 190 times, on October 5, 1971, the 191st experiment using low-temperature extraction with ether as the extract ingredient was successful in treating malaria.[20]

REFERENCES

1. Chung Yao Chih (Chinese Herbal Drugs), *Chinese Academy of Medical Sciences, Institute of Pharmacology*, Vol. 1, Jen Min Wei Sheng Press, Peking, 1959.
2. Ch'ang Yung Chung Yao (Commonly Used Chinese Drugs), *Ch'eng Tu Chung I Hsueh Yuan*, Shanghai Jen Min Wei Sheng Press, Shanghai, 1973.
3. Bowers, J. Z., *Chemtech*, August 1974, p. 458.
4. National Academy of Sciences. *Herbal Pharmacology in the People's Republic of China,* Botanical Medicine, Mary Ann Liebert Inc., Washington, D.C., 1975, p. 12. https://www.abebooks.com/9780309024389/Herbal-Pharmacology-Peoples-Republic-China-0309024382/plp
5. Veith, I., *J. Am. Med. Assoc.*, 228: 1577, 1974.
6. Huard, P. and M. Wong, *Chinese Medicine*, McGraw-Hill, New York, 1973.
7. Veith, I., *Huang Ti Nei Ching Su Wen (The Yellow Emperor's Classic of International Medicine)* (translation), University of California Press, Berkeley, 1966.
8. Smith, A. J. *Brit. Med. J.*, 2: 367, 1974.
9. Claus, E. P. et al., *Pharmacognosy*, Lea and Febiger, Philadelphia, 1970.
10. Martin, E. W. and E. F. Cook (Eds.), *Remington's Practice of Pharmacy*, 11th ed., Mack Publishing, Easton, PA, 1956, pp. 7–10.
11. Hsu, K.-C. and Chao, S.-H., *Pharmacognosy*, Hong Kong Wei Sheng Press, Hong Kong, 1971, pp. 1–34.
12. Hsu, K.-C. et al., *Yao Ts'ai Hsueh*, Jen Min Wei Sheng Press, Peking, 1961.
13. Horn, J. S., *Away With All Pests*, Monthly Review Press, New York, 1971.
14. Mao, T.-T., *Instruction on Public Health Work*, People's Health Press, June 3, 1967, p. 9.
15. Bowers, J. Z. and E. F. Purcell (Eds.), *Medicine and Society in China*, William F. Fell, Philadelphia, 1974.
16. Li, C. P., Chinese Herbal Medicine, U.S. Department of Health, Education, and Welfare, Publication No. (NIH) 75-732, 1974.
17. Wright, R. A., *The New York Times*, September 13, 1972.
18. Astin, J. A. *JAMA*, 279(19): 1548, 1553, 1998.
19. NCCAM, About NCCAM. General Information, 2002, http://NCCAM.NIH.gov/an/general4/15/2002
20. World Journal Today, October 6, 2015.
21. Ch'uo. Herbal Medicine *Dictionary*, vol. 7, p. 1228, Jiang Su New Medical College, China, 1975.

2 Ginseng in China

For centuries, the Chinese were—and they still are—the world's leading ginseng users. They value the root as a medicinal, a sort of curative charm. To the average Chinese, the value of a piece of wild ginseng root means more than gold or silver. Some authorities believe that ginseng's chief attraction for the Chinese lies in the root form, which branches out and resembles the human figure. As a matter of fact, the word *ginseng* or *Ren-shen* stems from two Chinese characters meaning *man* and *body*. The Chinese name of *ginseng* is often translated in Western works as *man-shaped root* or *manroot*. However, the botanist Professor S. Y. Hu of the Arnold Arboretum, Harvard University, prefers the term of *man-essence* for ginseng, since, according to Hu, ginseng was derived, on one hand, from the fanciful resemblance of the root to the human body, and, on the other hand, from the belief that this root represents the essence of the elixir of the earth crystallized in a human form. Because of this, it carries the nickname of *spirit of the earth*, and sick people are relieved from their illnesses after taking the root. Hence, traditionally, the more the root resembles the human figure, the more potent its healing properties are, and the greater is its worth.

For nearly 40 centuries, ginseng has been universally used in Chinese medicine as the most respected and superb health-maintaining tonic remedy. It has been prescribed repeatedly by doctors for some 40 different types of illnesses in China as well as in other Asian countries. Although its mechanism remains secret, the mysterious healing power of the ginseng root is indeed a miraculous and factual truth. To the average Chinese, ginseng means medicine par excellence (Figure 2.1).[1,2]

A great deal of interest in ginseng was generated in the mid-nineteenth century in the Western world. Scientific studies on the "man-essence" were initiated in Europe, particularly in France, Germany, and the Soviet Union. Recently, Japanese and Soviet scientists, after years of laboratory and clinical investigation, have confirmed that ginseng is capable of building up vitality and physical resistance, strengthening the organism and the endocrine systems, thus overcoming illness and maintaining homeostasis. Is it not amazing that it took nearly 50 years of endeavor to prove that the Chinese doctors' claims about ginseng's activities were basically correct?[3,4]

PANAX GINSENG: THE MOST PRECIOUS HERBAL MEDICINE

There are several varieties of ginseng in the world, the one best known being *Panax ginseng* C.A. Meyer, which is also called *Panax schinseng* Need, of the family Araliaceae. Meyer, who was a Russian, named the plants in 1843. The generic name of *Panax* is derived from the Greek *pan* meaning "all," and *akes* (άχέουxι), meaning "cure" or "heal." Accordingly, the word *Panax* means "cure-all," "all-healing," or "panacea." The word *ginseng*, or *schinseng*, or more correctly *Jen-shen* or *ren-shen* in

人参 （五加科）

Panax ginseng. From Handbook of Commonly Used Herbs in North China, *The People's Hygiene Publishing House, Perking 1971.*

FIGURE 2.1   Panax ginseng (Chinese ginseng).

Chinese, may mean ambiguously "human body," since the terms *gin, schin, Jen,* and *ren* all mean "man"; the terms *seng* and *shen* sound identical to "body" in Chinese.[5]

Panax ginseng is a perennial herb indigenous to the mountainous forests of Eastern Asia, particularly Eastern Manchuria (Liao-tung area), North China, Korea, and the maritime area of Siberia. Ginseng is a very long-lived plant. Chinese *Materia Medica* books had claimed that only the aged ginseng root, especially the hundred- or even thousand-year-old root, gives the most potent healing power. Scientifically, this may be unfounded, but ginseng's long life is a fact. Not too long ago, the Soviet botanist Grushvitzky found a Chinese mountain ginseng in the Manchuria forest, and after careful examination, the root was determined to be at least 400 years old.[6]

The Chinese believed that the best ginseng in China was the mountain-type ginseng grown only in the "Ch'ang-pai Shan" (long white mountains) area, which lies east of Liao-ning and south of Kirin Provinces in Manchuria. In the old days, this particular type of top-quality ginseng was extremely scarce as a result of its being difficult to collect; thus, the price of it was extravagant. This particular ginseng was called *Liao-tung shen*, or *Manchurian ginseng*.

Many mountain areas in North China, however, also produce ginseng, but not *Panax ginseng*. The well-known places are Shan-si, Hopei, Shen-si, and other provinces in northern and central China. Ginseng grown in Shang-tang (now Lu-an-fu) of Shan-si, was called *Tang-shen*, and in *Tah-Kwan Pen-ts'ao*, and many famous herb doctors claimed this to be the original and *genuine* Chinese ginseng with supreme healing power. Until the seventh century, both *Tang-shen* and *Panax ginseng* (Manchurian ginseng) were used widely in medicine. The original *Tang-shen* does belong to Araliaceae as can be seen from what is described in *Tah-Kwan Pen-ts'ao*.[7] However, up until the Ch'ing dynasty, the *Tang-shen* plant was classified into Campanulaceae, but no doubt the original Tang-shen has become extinct.

Three kinds of ginseng roots have been used in Chinese medicine: mountain or wild, transplanted, and cultivated ginseng. Wild ginseng grows naturally in the mountains, is the most expensive, and is supposedly the most effective. Transplanted ginseng is grown from young plants moved from the mountains to farms. Cultivated ginseng is the ginseng plant grown on ginseng farms after seeding. A Chinese ginseng expert, Li Chun-chin, however, classified Chinese ginseng into six grades according to the source and age of the plant: old mountain type (over 200-year-old ginseng root that can only be found in Ch'ang-Pai-shen and has been praised as the most superb quality); mountain type (a few to 50-year-old root); eradicate type (ginseng plant that was buried and revived in the mountains); transplanted type, Shih-chu ginseng (grown particularly in the Kuan-hsun area); and cultivated ginseng. The experienced ginseng dealer can easily distinguish whether the ginseng is a mountain or cultivated root and can approximate the age by examining the number of rings on the surface of the root itself and the above-ground part of the plant. Mountain ginseng was very scarce even in the old days. It was said that one *catty* of Manchurian cultivated ginseng was sold at a price of 2–20 *liang* of silver or gold, while a *catty* of the aged Manchurian mountain ginseng was sold at a few thousand *liang* of silver or gold (one *catty* is about 1.1 pound, or 500 grams; one *liang* is about 36 grams). At such extravagant prices, genuine ginseng could hardly be afforded by the common people for medicinal purposes. Naturally, only the rich, members of the emperor's families, and high officials benefited from the precious root.[5]

It is difficult to know exactly when and where the Chinese started the transplantation of or growing of ginseng on farms. However, the cultivation of ginseng appeared in the poems of the T'ang and Sung dynasties. More clearly,00 the *Pen-ts'ao Kang-mu* recorded the method for raising ginseng on plantations. Thus, it may be correct that ginseng cultivation in China started as early as in the fifteenth century, but large-scale ginseng cultivation in Kirin province did not start until the eighteenth century.

In Manchuria, the counties of Kuan-hsum and Juan-jen of Liaoning province, An-tu, Tun-hua, Chia-an, Fu-sung, Lin-chiang, and Wang-ch'ing of Kirin province, and Cheng-an-li of Hei-lung-kiang province are the chief areas producing wild as well as cultivated Chinese ginseng. Fu-sung is also a well-known area for cultivated ginseng. It is said that Fu-sung alone produces about 60% of the total ginseng in China. As a result of the constantly increasing consumption of ginseng in China, with little increase in production, the ginseng shortage remained as such for thousands of years, and China has never put great effort into cultivating on a large scale to meet her demand. Only since 1950 has the People's Republic of China put great effort into producing the wonder root under government supervision. The most suitable conditions for growing ginseng have been under extensive study at the experimental station of She-Baa-Wang, outside of Beijing.[8]

DESCRIPTION OF GINSENG

The earliest full description of the Chinese ginseng plant was given by Su Sung, in the eleventh century. In his herbal dispensatory book *T'u-ching Pen-ts'ao* ginseng was described as follows:

> Ginseng is grown in moist and shaded forest of Chia tree (Chia is *Tilis murensis* or *Tilia manshuria* species). The Chia tree is a broad-leaved tree, thus providing good shade. The dried ginseng seeds were planted in October as for vegetable seeds. The ginseng dies each year, and sprouts emerge in the Spring. The young plant is about 3–4 inches high with only one stem which carries five small, parted, palmate-shaped leaves. Four to five years later, it grows two stems, but still no flower stem. Until it is ten years old, it has three, later four stalks all rising from the same center, each carrying five leaflets, three large and two small. At the center of the apex of the elder plant is a flower stalk. Unlike the majority of mountain plants, ginseng has only one flower stalk which blossoms in the late spring. The ginseng blossom is small and umbrella-shaped, the size of a chestnut, and purple-white in color. It bears 7–8 seeds of soybean size after autumn. The seeds are green in color when fresh, but turn into red when ripe and drop into the ground automatically.

A Jesuit missionary in China, Fr. Jartoux, was perhaps the first Westerner who witnessed the gathering and use of ginseng in Manchuria. In a letter from Beijing dated April 12, 1711, addressed to the Procurator General of the Mission of India and China, Fr. Jartoux furnished a detailed description of *Panax ginseng* that he observed in Manchuria.[8]

CHARACTERISTICS OF GINSENG PLANT

Ginseng is a very slow-growing plant. The root of ginseng is collected only at a certain age of the plant and at a certain season. For example, the root is firm if it is collected in September or early October and soft if it is collected in spring or summer. The outside skin of the fresh root is yellow, but inside the skin is white. The length and diameter of the root vary with the age of the plant. It can be a few inches (a few years old root) to a foot (after 10 years). The fresh root tastes slightly bitter and somewhat sweet, and has a typical ginseng aroma.

The well-known botanist, Professor Baranov of Arnold Arboretum, Harvard University, published an article on morphology, cultivation, and use of ginseng.[9] The peculiar behavior of ginseng was described as follows:

Not just mysterious, the ginseng root is also peculiar. It belongs to the category of contractile roots, and is an important part of a mechanism which ensures proper position for the "regeneration bud" of ginseng. The root of ginseng is normally crowned with an underground stem or vertical *rhizome* called the "neck." The rhizome grows upwards and increases yearly in length at the upper end. The regeneration bud is formed at the apex of the rhizome, and must necessarily find itself just at the soil-level. If the growth of the rhizome continues uncontrolled, it will finally emerge from ground and bear regeneration buds above the surface. Such a situation would be a disadvantage for the plant. Therefore, to counter-balance the lengthening of the rhizome, the ginseng root shrinks yearly at the same rate at which the rhizome grows upward, and pulls the plant downward. As a result of their mutually opposed movements, the tip of the rhizome with the regeneration bud finds itself always exactly at the soil level. This interesting morphological and biological peculiarities of ginseng plant was discovered only a few years ago.

PREPARATION OF GINSENG ROOT

After the ginseng root is dug in late September or October, the root has to be treated by one of the traditional methods to make commercially acceptable and easily preserved products. There are at least six kinds of Chinese ginseng root products, prepared by six different methods.[10]

Sheng-shai shen (plain, dried ginseng) is made initially by washing and cleaning the fresh root and carefully freeing it of adhering soil without scraping the outside skin and small roots. The clean root is hung up in the air and dried completely. It is usually yellowish and slightly gray in color, with characteristic circular markings that remain untouched after this process.

Pai-kan shen (dried white ginseng) is made by cutting off the small roots and the hairy roots, scraping the outside skin from them, and drying completely in the sun. The root thus prepared is very white and smooth.

Ta-li shen (dried ginseng root) is made after carefully freeing the root from the soil, cutting off the small roots, the branches, and the hairy roots, then boiling briefly and drying in the sun or above a charcoal fire to complete dryness.

T'ang shen (sweetened ginseng) is made after cleaning the root, briefly boiling and puncturing the root, and then putting it in syrup for about 24 hours. The sweetened, sugar-preserved ginseng root then is dried in the sun to complete dryness. The root thus prepared is white and sweet.

Hung shen (red ginseng) is made after carefully freeing the root of soil with a brush, cutting off the hairy and branched roots, and brushing the skin until it looks white. Then the root is put in a steamer to steam for about 3 hours. Then it is dried in the sun to complete dryness. The root thus prepared is semitransparent and reddish-brown in color. This is the most popular ginseng root preparation.

Shen-shu (ginseng tails) are the small ginseng roots and hairy roots that can also be made into different forms, white or red, according to the methods for the main roots.

How to Keep Ginseng Roots

Ginseng roots should be kept in a dry and dark place. Since it is subject to being worm-eaten and very liable to be attacked by insects, the root must be kept in a hermetically sealed container or jar.

Adulteration

Since the beginning of the ginseng trade, the adulterated roots were found present and fraudulently substituted for genuine ginseng on the market. The roots of several campanulaceous plants, such as *Sha-shen* (*Adenophora verticillata*), *Chi-ni* (*Adenophora remotifolia*), *Ti-ni* (*Adenophora tracheloides*), and *Chieh-keng* (*Platycodon grandiflorum*), bear close resemblance to those of ginseng and are most frequently the adulterated species. These plants and their overall pharmacologic properties, however, are different from ginseng. For example, the root of *Sha-shen* is more spongy and slightly cooling and demulcent. *Sha-shen* is used in Chinese medicine as an expectorant.[5] The root *Chi-ni* resembles that of *Sha-shen*. The root of *Chi-ni* is sweet and cooling. *Chi-ni* and *Chieh-keng* are botanically alike, and their medicinal values are also as expectorants.[11]

HISTORY OF THE MANROOT

The first august sovereign and the first unifier of China, Shih Huang Ti of the Ch'in dynasty (221–209 BC), sought to create a durable, centralized, and unified empire that would last ten thousands of years. Legalism became the state orthodoxy. The emperor decreed the burning of all political writings and undesirable books. Toward the always dangerous enemies from the North and the West, the emperor built up the Great Wall of China, a solid, dense barrier from the sea to the west desert along the mountains that marked the northern boundaries of ancient China. The emperor also wanted longevity and immortality. Without losing any time, the emperor dispatched a group of three thousand young men and three thousand young women headed by Hsu Fu to the mountains in a remote eastern place that, he once heard, produced the "divine herb." This place, called Pong-lai, was later confirmed to be Japan. The divine herb that the emperor desired was said to produce rejuvenation and longevity. The supreme desire in human life is unreachable by the common man. Unfortunately, this group of immortal herb collectors never returned to China, and the great emperor never had his desire of immortality fulfilled. Later people speculated the "miracle herb" could be ginseng.[12]

During the Sui dynasty (AD 581–601) in the reign of Emperor Wen, the history book of *Kwang-wu Hsing-chi* records the man-shaped ginseng cry story.[1] At Shang-tang (now Lu-an of Shan-si province), at the back of a family's home, each night the imploring voice of a man was heard and nothing was found when searches were made for the source of the noise. At a distance of about a *li* (a *li* is about three-quarters of a mile) from the house, a remarkable plant was seen. A root was secured after digging into the ground to the depth of about 5 feet. The root had the shape of a man with four extremities resembling legs and arms. After this, the noise ceased, and so it was said that this plant had caused the crying out in the night with a man's voice. Thus, the root was called "spirit of the ground." The mysterious nature of the man-shaped root may thus have been started.[12]

Some of the dry ginseng roots indeed vaguely resemble the human figure, with the head (the above-ground part) on the top, and the arms and legs on the upper and lower parts of the root, the side roots). It was said that the "man-shaped" root could be made artificially by traders or hunters. The man-shaped root usually is worth much more than the random-shaped root, which normally occurs.

In the old Chinese rural society, under the strong influence of Taoism and Buddhism, the objects encountered by the ancient Chinese people were veiled with a great deal of mystery as to their color, shape, size, smell, taste, etc., most especially so, their use as food and medicine. Wild (mountain) ginseng was believed to give better healing power than cultivated ginseng. The man-shaped ginseng root was regarded as the "divine" or "immortal" herb; after taking the root people were thought to become long-lived or "immortal."

There were quite a number of legendary stories about ginseng in ancient China. Because of the superstitious nature of the Chinese, one finds recorded in the *Pen-ts'ao Kang-mu* at least 10 names, strangely enough, to describe the ginseng plant and its root. *Kuei-kai*, translated as "ghost umbrella," means the ginseng plant is a shade (dark)-lover, and is always hiding itself from the sunshine.

Shen-ts'ao, translated as "divine herb," means the ginseng plant has panacea power over disease.

Tu-ching, translated as "the spirit of the earth," and *Ti-ching*, translated as "the spirit of the ground," mean the mysterious nature of the ginseng root, which may be equivalent to "ghost" or "spirit of the earth."

Shueh-shen and *Hung-shen*, translated as "blood-root" and "yellow root," mean the ginseng root is tonic to the spleen (yellow in nature), which in turn produces blood.

Jen-wei, *Jen-hsien*, *Hai-yu*, and *Chou-mian-huan-tan* are strange names without any clear meaning.

It is difficult to know exactly when ginseng was first used in Chinese medicine. *Shen-nung Pen-ts'ao Ching* gives descriptions of ginseng and classifies it as a superior and nontoxic drug. *Ming-I, Pieh-lu* has more entries about the uses of ginseng. In Chang Chung-ching's famous medical book *The Treatise of Fevers*, written in the Later Han dynasty (circa AD 195), 21 out of a total of 113 prescriptions contain ginseng for different ailments. Accordingly, there is no doubt that ginseng had been used as medicine in China as early as the second century,[5] or about 18 years ago.

During the Epoch of Three Kingdoms, the medicinal consumption of ginseng was much increased. The great surgeon Wa-t'o, discovered the use of ginseng in a preparation for nasal and internal bleeding.

According to another history book, *Wei-shu*, Koguryo (Kaoli), a kingdom in the northern Korean Peninsula, now North Korea and a part of Manchuria, started a diplomatic relationship with China in the third century (Wei period, 220–264). Within 100 years, more than 92 diplomatic mission trips were made by the Koguryoan envoys. On each trip, the envoy brought to the Chinese emperor ginseng roots and other valuable gifts from Koguryo, and on return brought back Chinese silk, culture, and medicine. During the T'ang dynasty (618–905), the envoy from Silla (another kingdom in the southeastern part of the Korean Peninsula) made quite a number of diplomatic mission trips to China, and on five trips brought ginseng from Silla to China. The Silla ginseng roots were man-shaped, about 1 foot in length, wrapped in red silk, and contained in wooden boxes. However, another history book, *San-kuo Shih-Chi*, recorded that once an envoy from Silla brought to the Emperor The of the T'ang dynasty a giant 9-foot ginseng root, but the Chinese emperor refused to receive the gift because it was not a genuine ginseng root.[12] As a result of increased communication between China and the Korean Peninsula, more ginseng roots were sold to China from Silla and Koguryo, and later the third kingdom, Packche (located in the southwestern part of the Korean Peninsula), also exported ginseng to China. In other words, since the third century, the Chinese have been keen users of Korean ginseng.

The book *Ming-I Peih-lu* gives a brief description of the Koguryoan's ginseng-hunting story. The diggers, as a rule, first prayed to their gods of the mountains for safety and good luck. After worship, as a group, they started their trip and entered the mountains with enough food, warm clothing, and arms. Conceivably, the ginseng-hunting trip was fearful and dangerous, and quite often they encountered icy weather and even cruel wild animals, and no one knew if they would return alive. The people of Koguryo also have beautiful hymns to praise their mysterious divine root, which they used to call *Hsien-ts'ao*, meaning "immortal herb." The following hymn tells how the Koguryoans searched for mountain ginseng:

> The people of Koguryo (Kao-li) praise their ginseng. The one having the three stems, five palmate leaflets, facing the shade and away from the sun is ginseng. To search for ginseng, one must discover *Tilia*, for *Tilia manshuria* always accompanies ginseng.

During the Sung dynasty, ginseng fell into seriously short supply, and the quality of ginseng and adulterated ginseng became quite a problem. One very simple method, which may be the first pharmacologic test of ginseng in human history, is that according to Su Sung's *Tu-ching Pen-ts'ao*: "In order to test for the true ginseng, two persons walk together, one with a piece of ginseng root in his mouth and the other with his mouth empty. If at the end of 3–5 *li*, the one with ginseng in his mouth does not feel himself tired, while the other is out of breath, the ginseng is genuine ginseng root."[1]

The natural ginseng plants growing in Hopei, Shan-si, and Shen-si provinces were nearly extinguished up to the Ming dynasty. Manchurian and Korean (Silla, Packche, and Koguryo) ginseng then became the only supplies available for the great demand. Korea began to export cultivated ginseng to China during the Ch'ing dynasty at about 100,000 *catties* annually. Starting in 1875 China began to import American ginseng from the United States at about 60,000 pounds annually. As a result of the *sang* (American ginseng) boom between 1895 and 1904, an extravagant fortune was made by the ginseng diggers and traders in the United States.[13]

VIRTUES OF GINSENG IN CHINESE MEDICINE

Based on accumulated knowledge and experience acquired in the treatment of patients during thousands of generations, doctors in the Orient, particularly in

China, do believe the healing power of ginseng. Nobody really knows how many people have been treated with the manroot in the last 20 centuries. The miraculous power of ginseng, no doubt, arises from its many therapeutically effective principles. According to classical Chinese *Materia Medica* books, ginseng possesses the following medicinal properties:

- A mild stimulant to the heart, nervous system, and digestive organs
- A tonic to an impaired constitution, to add spirit, to increase digestive juices, to speed up recovery after a long and serious illness and after a surgical operation[5]

Ginseng is an agent that increases digestion after oral administration. It is absorbed in the small intestine, where it enters the bloodstream. It promotes blood circulation and new blood formation, thus invigorating your spirit and strengthening your body. The principal usefulness of ginseng is its *tonic* effect. It is absolutely essential to those who are suffering from diseases of consumption, neurasthenia, and its related dizziness and headache, impotence and related loss of sexual potency, impaired kidney and uterus functions, and illness due to extensive daily physical and mental activities or stresses.

In the history of Chinese medicine, the *Shen-nung Pen-ts'ao ching* was the first official medical book that stated the pharmacologic virtues of ginseng.[14] The following statements are still true and universally accepted:

- Tonic to the five viscera
- Quieting the spirits
- Establishing the soul
- Allaying fear
- Expelling evil effluvia
- Opening up the heart and brightening the eyes
- Benefiting the understanding
- Invigorating the body and prolonging life, if it is taken constantly

Other famous Chinese medical treatises (see Chapter 1) such as *Ming-I Pieh-lu, Hai-yao Pen-ts'ao, Yao-hsing Pen-ts'ao,* and the most famous Chinese doctor, Chang Chung-ching's prescription books called *Wai-t'ai Mi-yao, Chia-yio Tu Ching Pen-ts'ao, Pen-ts'ao meng-ch'uan, Sih's Medical Compendium,* and Li, shin-chen's *Pen-ts'ao Kang-mu* also record the medicinal properties and different uses of ginseng.[5]

Ming-I Pieh-lu described ginseng as an effective drug for the following:

- Chronic gastrointestinal disability
- Gastric and intestinal pain as a result of swelling and gas
- Dyspepsia (impaired digestion)
- Difficulty in respiration
- Acute gastritis and enteritis, and vomiting
- Increase in digestive functions
- Elimination of thirst and polyuria as a result of diabetes

- A heart tonic, and strengthening of blood circulation
- Inflammation and swelling
- Increase in memory

In the *Hai-yao Pen-ts'ao* ginseng was recorded to be effective in the treatment of thirst (as a result of diabetes), mental nervousness, loss of body fluid (as a result of fatigue), and inhibition of hyperchlorhydria (excess stomach acid).

Most important of all, the famous pharmacologist Li shih-chen, after repeated testing and experimenting, listed the activities and indications of ginseng in his world-famous *Pen-ts'ao Kang-mu* as follows:

- All forms of debility of man and woman
- Various types of severe dyspepsia (impaired digestion)
- Continued fever and cold perspiration
- Drowsiness and headache
- Persistent vomiting of pregnant woman
- Chronic malaria
- Exhausting discharge, polyuria, internal injuries
- Apoplexy (loss of consciousness and sensation)
- Sunstroke and paralysis
- Hematemesis and menorrhagia (excessive menstrual flow)
- Bleeding in feces and urine
- Hemorrhage and puerperal diseases

The Chinese doctors, however, also warned people *not* to abuse ginseng and *not* to take ginseng in large quantities (over 1 *liang*) without a prudent diagnosis by an experienced physician. Since ginseng is *not* a *placebo* but a potent remedy, its effectiveness depends on its current use. It is perfectly all right if you take ginseng tea or small amounts of extract every day for health-maintaining purposes, but if you are really sick, you ought not to take for granted that ginseng will be effective in the treatment of certain unknown diseases, in which case it may be harmful or even hazardous.[5]

CLASSICAL GINSENG PREPARATIONS AND SECRET PRESCRIPTIONS

Ginseng can be used alone or, as a rule, with several other ingredients in order to give a multitherapeutic or synergistic effect of healing. As mentioned in the last chapter, the "complex-remedy therapy" is the characteristic and standard practice of Chinese medicine. Ginseng extract, decoction, tincture, powder, and pills are the most commonly used ginseng preparations. According to the prescription, the drugs are weighed, mixed, and a decoction or other form of preparation is made by the druggist. The dose and how often the patient has to take the preparation are also indicated in the prescription. The following are examples that explain what and how the classical Chinese ginseng preparations are usually made in typical Chinese drugstores.[15,16]

Ginseng Extract (*Jen-shen Kao*) This is a watery extract made by fractionally decocting ginseng root in water and evaporating the extract to a thick liquid in a silver or earthenware pot. It was told that ginseng should never be cooked in a metal pot, except silver. The extract is normally a dark yellow or brown liquid with typical ginseng bittersweet taste and aroma. This extract is kept in a porcelain container.

According to the Chinese *Materia Medica* book, *Pen-ts'ao Kang-mu*, the previously discussed ginseng extract is made in the following manner: In a silver or earthenware pot, add 10 *liang* of ground ginseng, pour into the pot 20 wine-cups of water, and soak for a little while. Then boil the infusion over a gentle flame until half of the water is evaporated. The mixture is strained (filtered) through two layers of cheese-gauze. Then put the filtrate aside. The solid portion is cooked again in the pot with 10 wine-cups of water boiling the infusion until half of the volume is reached, and then filter again. Combine the two portions of the filtrate and return them into the pot, and then boil for some time until this liquid is very thick. Store this ginseng extract in a covered porcelain container or jar.

If the extract is used for curing disease, usually it is taken once or twice a day before meals, the dose depending on the requirement of the patient. For tonic purposes the above extract (containing 10 *liang* of ginseng) can be divided into 30–40 doses, which is the usual dose of ginseng for health-maintaining tonic effect.

Ginseng Decoction (*Jen-shen t'ang*) Most Chinese medicines are administered in the form of a decoction, which is the simplest way to give medicine. It is made similarly by boiling the drug ingredients in an earthenware pot with water over a gentle flame for usually 1–2 hours. Unlike the extract, no further thickening is required. The decoction can be filtered, and only the liquid portion is used for medicine. The entire decoction can be administered all at once or divided into several doses depending on the amount of drug cooked and the prescription.

Many secret prescriptions of ginseng decoction are used in China, Korea, and Japan. The following are examples of the most popular ginseng formulations for tonic effect (Table 2.1).

Put the preceding ingredients into an earthenware pot, add water (about 2 cups), and bring to a boil over a gentle flame and keep boiling for some time. Strain the mixture, and use the filtrate as medicine. This particular preparation has been very widely used in China as a tonic for wasting, weakness, and tiredness (Table 2.2).

Put 3 *liang* of each of the preceding ingredients into an earthenware pot, and add about 8 pints of water and then bring to a boil. Boil over a gentle flame for some time or until 3 pints remain. Strain the mixture, and the liquid portion is used. Take 1 pint

TABLE 2.1
Szu-Chun-tze T'ang (Gentlemen's Ginseng Decoction)

Ginseng	1 *chien* (3.6 g)
Pai-chu	2 *chien* (7.2 g)
Fu-ling (Indian bread)	1 *chien* (3.6 g)
Licorice root	0.5 *chien* (1.8 g)
Fresh ginger root	2 slices (2 g)
Red dates	1 each

TABLE 2.2
Chih-Chung T'ang (Resolvent Ginseng Decoction)

Ginseng	3 *liang* (150 g)
Pai-chu	3 *liang*
Dried ginger	3 *liang*
Licorice root	3 *liang*

each time, three times a day. It is a wonderful preparation for illness of the lungs, spleen, heart, and stomach (Table 2.3).[1]

Add about 6 pints of water to the preceding drug ingredients in a pot and bring to a boil, boiling until the liquid is reduced to about 2.5 pints. This decoction can be divided into four doses. It has been used as a general-purpose tonic for weakness (Table 2.4).[1]

In a similar manner, the preceding drug ingredients are made into a decoction. Taken before meals, it serves as a potent tonic for good health and longevity.[15]

A decoction used for restorative purposes is made with ginseng extract, orange peel tincture, and ginger juice, each of suitable amounts, taken before meals.

TABLE 2.3
Sze-Shuen T'ang (Four-Drug Decoction)

Ginseng or ginseng extract	2 *liang* (100 g)
Licorice root	2 *liang*
Dried ginger	2 *liang*
Fu-tze-p'ao	2 *liang*

TABLE 2.4
Ch'ang-Shou-T'ang (Decoction for Longevity)

Ginseng root	
Huan-chi (Yellow vetch)	
Pai-chu	1 *chien* (3.6 g)
Tu-chung	
Niu-hsi (Ox knee)	
White peony	
Licorice root	0.6 *chien*
Wu-wei-tze	12 *pieces*
Shu-ti-huang	2 *chien* (7.2 g)
Red dates	1 each

Another decoction used for restorative purposes is made with ginseng extract, orange peel tincture, and honey, each of a suitable amount. Mix them together, and drink it before going to bed.

Ginseng Tincture (*Ren-shen chiu*) Traditionally, tinctures are made by macerating the ground drug in a mixture of rice and leaven during the process of fermentation for producing spirit. However, the modern method of making tincture can be achieved by extracting the drug with wine or spirit to give an alcoholic tincture. The ginseng tincture preparation has been made in Chinese dispensatories, but it is not as popular as ginseng decoctions. However, the ginseng tincture mixed with tincture of *Kou-chi* (*Lycium Chinense*) and/or tincture of *Lu-jung* (*Moaochasme savatieri*) has been used for tonic purposes, especially, for sexual debility and impotence in males.[5]

Ginseng Powder (*Ren-shen San*) The powdered ginseng root or the dried ginseng extract can be used as tea or soup, to drink alone or with another substance such as sugar or honey. This is a tonic of choice for those who have peptic ulcers and are unable to take ordinary drinks of coffee or tea.

Ginseng Pills (*Ren-shen Wan*) In Chinese medicine, pills have been widely used and are a favorite solid preparation for exhibiting drugs without exposing disagreeable tastes to the patient. Pills are generally divided into *Wan*, *Tan*, and *San*, made in all sizes from that of a millet seed to that of a pigeon's egg. Pills are usually coated with some typical coating materials to mask the bitter taste. Rice flour and honey are the most widely used fillers in making pills. Among hundreds of pills preparations in Chinese medicine, the following are some of the most popular ginseng pill formulations.

Tsao-shen Wan (Date Ginseng Pills)

These are made of large Chinese red dates and ginseng extract. They are said to be useful in strengthening the respirative organ.

Chen-Jen Pao-Ming Tan (Sour-Date Ginseng Pills)

These are made of *Suan-tsao* (*zizyphus jujuba*), ginseng, *Fuling*, and *T'ien-men-tung*, three *chien* of each, soaked with wine or spirit for some time (3 days). Then the alcoholic extract is evaporated to dryness and made into pills. Take one-fourth of the total pills before going to bed. They are used for increased sexual potency.[16]

Sen-shen Kwie-pi Wan (Ginseng Cinnamon Pills)

Mix powdered ginseng, *Jou-Kuei* (fleshy cinnamon), *Mai-men-tung*, *Wu-wei-tzu*, and other excipients to make pills. It is said that this preparation has been used traditionally in the Orient as an aphrodisiac.[5]

OTHER MIRACULOUS ROOTS AS GINSENG SUBSTITUTES

There are at least eight medicinal plants that are cousins of *P. ginseng*. Because the roots of these plants bear close resemblance in action to ginseng, they also carry the general name of *shen,* and they have been widely used in the Orient for ginseng substitute. Each of them occupies a particular place in Chinese medicine.[1,5] They are *Sha-shen, Hsuan-shen, Tan-shen, Kú-shen, Tzu-shen, Tai-tzu-shen, Tang-shen,* and *Jen-shen-san-ch'i.* Though these roots are less well known than ginseng, they are equally useful medically, and their prices are much lower.

It was recorded as early as in *Min-I-Pieh-Lu* that five miraculous remedies, more correctly called *Wu-shen,* were known in Chinese medicine. The *Wu-shen* are *Ren-shen,* (ginseng), *Sha-shen, Hsuan-shen, Tan-shen,* and *Kú-shen.* None of these five remedies belongs to the same family botanically. However, the five slightly different plants were grouped together by Li Shih-chen. Thus, the *Wu-shen* are *Ren-shen,* (ginseng), *Sha-shen, Hsuan-shen, Tan-shen,* and *Tzu-shen.*

It is very interesting to note that the dried roots of each of these species carry a particular color, and thus it was assumed that each of the plants may act specifically on each of the principal five viscera. Certainly this complies with ancient Chinese medical thinking. *Ren-shen* has been claimed to act chiefly on the spleen, the center of life. Being yellow in color, *Ren-shen* has been called *Huang* (yellow)-*shen; Sha-shen,* grown in sand-soil, is much whiter than *Ren-shen* (ginseng), and is also called *Pai* (white)-*shen.* It acts principally on the lungs; *Hsuan-shen* is also called *Hei* (black)-*shen.* Because its root, stem, and seeds are dark or nearly black, it is supposed to act on the kidney; *Tan-shen,* being red in color, is also called *Ch'ih* (red)-*shen.* It acts on the heart and the blood. *Tan-shen* has been highly recommended in all blood difficulties such as hemorrhages, menstrual disorders, and miscarriages. *Tzu-shen,* also called *Mou-meng,* is purple in color and acts chiefly on the liver. It has been prescribed for kidney and blood disorders.[17] The truth of these medicinal claims needs scientific confirmation, and little data are available at present.

SHA-SHEN

Sha-shen has also been called *Yang-ru* (goat milk) in addition to *Pai* (white)-*shen.* Its roots give milky white juices, and it is much whiter than the *Ren-shen* root. Botanically, *Sha-shen* is *Adenophora verticillate* of the family Campanulaceae. The best *Sha-shen* was from the Hwa-shan area in China. Natural and cultivated *Sha-shen* are now grown in the southeast and the north of *Sze-chwan* provinces of China. The root of *Sha-shen* tastes slightly bitter and cool. It has been used for pulmonary disorders, especially those attended by fever and cough, and it is used as a general tonic and restorative agent. *Ren-shen* and *Sha-shen* have been prescribed selectively for different symptoms. For lung disorders with burning, *Sha-shen* was used; for lung disorders with cooling, *Ren-shen* was used. It was said that *Sha-shen* is a tonic for the yin, while *Ren-shen* is a tonic for the yang of the five viscera.[17]

HSUAN-SHEN

Hsuan-shen has also been called *Yeh-chih-ma* in addition to *Hei* (black)-*shen.* Botanically it is called *Scrophularia Oldhami,* Oliv. It blossoms in March, grows about 4–5 feet high, and has a slender stem that resembles that of ginseng. It has opposite leaves, long and serrated, resembling those of wild sesame. It bears black seeds and also greenish-blue or white flowers in August. The root and stem of *Hsuan-shen* have a fishy odor.

The root tastes slightly bitter and cooling. It has been used as a tonic, restorative, and diuretic. It also has been used in stomach and intestinal fevers, malaria, extreme thirst, abscess, scrofulous glands, and galactorrhoea.[1,18]

TAN-SHEN

Tan-shen has also been called *Chu-ma* or *Beng-ma ts'ao* in addition to *Ch'ih* (red)-*shen*. Botanically, it is *Salvia miltiorrhiza* Bunge of the family Labiatae. It is now grown in the Ho-pei, Shan-tung, An-hwei, Szu-chẃan, and Kiang-su provinces of China.[1,19]

It sprouts in February and grows to about 1 foot high with opposite leaves that resemble those of peppermint leaves. It has purple blossoms starting in April and bears red fruit.

It has been prescribed for many blood disorders, particularly menstrual ailments, miscarriages, and hemorrhages. It stimulates blood circulation and the formation of blood cells. It has also been used for beriberi, joint pains and arthritic ailments, nervous insomnia, enlargement of the spleen, and hypertension.[20]

In animal health, *Tan-shen* has been used for horses, stimulating racing horses for prolonged activities and better performance.[1]

K'U-SHEN

K'u-shen has also been called *K'u-shih* or *K'u-kuo* (bitter bone), *k'u* meaning "bitter" in Chinese. Botanically, it is *Sophora angustifolia* of the family Leguminoseae. It is a very common plant in central China and Manchuria. It bears yellowish-white flowers, has a siliquaceous pod, and has a long yellowish and exceedingly bitter root. The best species is from the Ho-nan province of China.[21]

TZU-SHEN

Another name of *Tzu-shen* is *Mou-meng*. Botanically it is *Polygonum bistorta* of the family Polygonaceae. It has purple-white blossoms in May, and bears black seeds. It has a purple-black root with a bitter and cool taste.[1]

It has been used as a tonic, an antifebrile, a diuretic, and a laxative. It has also been prescribed for internal and external hemorrhages, amenorrhea, agus, and dysentery.[22]

TAI-TZU-SHEN

Botanically it is called *Pseudostellaria rhaphanorrhiza* Pax. of the Caryophyllaceae family. It has also been called *Tzu*(baby)-*shen*. It is produced mainly in the Kiang-su province of China. It has been used as a ginseng substitute for tonic action.[23]

TANG-SHEN

There are many varieties of *Tang-shen* in China, but the most important one is botanically *Condonopsis tangshen* Oliv. of the family Campanulaceae. It is produced mainly in the Shan-si, Shen-si, Ho-pei, and Kan-su provinces of China. It was used as a ginseng substitute.[23]

REFERENCES

1. Li, S.-C., *Pen-ts'ao Kang-mu*, Vol. 12 (original), Shang-wu Press, Shanghai, China, 1970.
2. Wallnöfner, H. and A. von Rottauscher, *Chinese Folk Medicine*, New American Library, New York, 1972, p. 44.
3. Brekhman, I. I. "Panax Ginseng," *Medgiz* (Leningrad), 182: 1957.
4. Brekhman, I. I. and I. V. Dardynov, *Ann Rev. Pharmacol.*, 9: 419, 1969.
5. *Chung-Kuo Yao-hsueh Ta-tzu-tien (The Encyclopedia of Chinese Herb Medicine)*, Vol. I. Hong Kong I-Yao Weisheng Press, Hong Kong, 1972.
6. Grushvitzky, I. V., *J. Bot.*, USSR, 44: 1694, 1959.
7. Wang, K. H., *Yao Hsueh T'ung Pao*, 1: 264, 1953.
8. Fournier, P., *La Nature* (Paris), 244, 1940.
9. Baranov, A., *Econ. Bot.*, 20: 403, 1966.
10. Hsu, K.-C. and S.-H. Chao, *Pharmacognosy*, Hong Kong Wei-sheng Press, Hong Kong, 1971, pp. 143–149.
11. Smith, F. P. and G. A. Stuart, *Chinese Medicinal Herbs*, Georgetown Press, San Francisco, 1973, pp. 15–17.
12. Tan, P.-Y., "Collection of Chinese Medicine Papers on Ginseng," in *Huang-han-I-hsueh ts'ung-shu*, Vol. 14, Chingyo Press, Taipei, 1970.
13. Williams, L. O., *Econ. Bot.*, 11: 344, 1957.
14. Smith, F. P. and G. A. Stuart, *Chinese Medicinal Herbs*, Georgetown Press, San Francisco, 1973, p. 304.
15. Imamura, T., *Ninjin Shinso*, Korean Ginseng Monopoly Bureau Press, Kae-Jo, 1923.
16. Imamura, T., "Medical and Pharmaceutical Aspects," in *Ninjin-shi, (History of Ginseng)*, Vol. 5, Shibunkaku, Kyoto, 1971, pp. 232–242.
17. Smith, F. P. and G. A. Stuart, *Chinese Medicinal Herbs*, Georgetown Press, San Francisco, 1973, pp. 301–304.
18. Smith, F. P. and G. A. Stuart, *Chinese Medicinal Herbs*, Georgetown Press, San Francisco, 1973, p. 400.
19. Hsu, K.-C. and S.-H. Chao, *Pharmacognosy*, Hong Kong Wei-sheng Press, Hong Kong, 1971, pp. 463–465.
20. Smith, F. P. and G. A. Stuart, *Chinese Medicinal Herbs*, Georgetown Press, San Francisco, 1973, p. 392.
21. Smith, F. P. and G. A. Stuart, *Chinese Medicinal Herbs*, Georgetown Press, San Francisco, 1973, p. 414.
22. Smith, F. P. and G. A. Stuart, *Chinese Medicinal Herbs*, Georgetown Press, San Francisco, 1973, p. 341.
23. Hsu, K.-C. and S.-H. Chao, *Pharmacognosy*, Hong Kong Wei-sheng Press, Hong Kong, 1971, pp. 460–461.

3 Ginseng in North America

The ginseng plant found in North America is commonly called American ginseng and is only slightly different from the one native to China. American ginseng is also called *sang, red berry,* and *five fingers*. Botanically, it is known as *Panax quinquefolius* Linn. of the Araliaceae family named by Linnaeus in 1753. It is a fleshy, rooted, perennial herbaceous plant.

Historically, it grew naturally on the slopes of ravines and in other shady but well-drained areas in hardwood forests from Quebec to Manitoba, and from Maine and Minnesota southward to the mountains of Georgia, Arkansas, and Louisiana. The properly dried roots were used as remedies. Wild mountain ginseng roots from the northern part of the United States, particularly in Wisconsin, Pennsylvania, and New York, are the most desirable, commercially favored products for exporting. These areas furnish roots of good size, weight, and shape, and they are generally considered the best breeding stock.[1] A related plant, the "dwarf ginseng," officially known as *Panax trifolium* Linn., is found from Nova Scotia to Wisconsin and Georgia. But this species is not desirable commercially.[2,3]

DISCOVERY STORY OF AMERICAN GINSENG

A French Jesuit priest, Father P. Jartoux (1669–1720), with the help of P. Regis, serving as special topographic advisers to the Chinese Emperor Kanghsi, arrived in Beijing in 1702. They were immediately sent to survey the remote lands in Manchuria, and on their journey they were able to visit the homeland of Chinese ginseng and experience firsthand the healing power of ginseng. Father Jartoux was the first who furnished a detailed description of the Chinese ginseng (*P. ginseng*) plant in a letter that arrived in Paris on April 12, 1711.[2] The missionary described the plant as "a Tartarian plant, called ginseng, with an account of its virtues in medicine." In 1714 the missionary published the ginseng story in the *Philosophical Transaction* of the Royal Society of London. The news soon stimulated serious interest in Europe in finding a miraculous panacea.

As a matter of fact, Chinese ginseng was not totally unknown to the Europeans, since back in the thirteenth century, Marco Polo, who found ginseng in general use throughout China, recorded it in his travel narrative in 1274 and made it known to other Europeans. He described the use of ginseng: "It was powdered, cooked, and used as a tea, syrup, or food condiment or even burned as incense in the sickroom."[4] Later, Dutch merchants brought ginseng roots from the Orient to Europe in 1616, but except from some curiosity by the seamen, no practical interest or attention was given it.

It was said that even the King of France chose Chinese ginseng instead of several other very valuable gifts from Asia offered by a Siamese ambassador in Paris in 1686, since it was believed at that time that ginseng was the most precious panacea in the world. In 1697, the French Academy of Sciences conducted a discussion of ginseng, and ginseng in France was used for asthma, stomach disorders, and to promote fertility in women.[5]

Ginseng then became a desirable plant, and people in France turned to it with great enthusiasm. Michael Sarrasin, the king's physician for Canada, appointed by Louis XIV, collected some plants in Canada that were suspected of being ginseng plants.

The Indians in Canada knew about the ginseng plant, and they called it *Oteeragweh*. Some of the plants collected near Quebec by the Indians were sent to Paris for examination.[6,7]

Father Jartoux's communication created a tremendous interest in the Western world, and suggested that the valuable root might be found in some countries, particularly French Canada where the forests and geographical environment are very close to those in Manchuria. Also fascinated about the ginseng business was a Jesuit missionary among the Iroquois Indians, Father Joseph Francis Lafitau, who also supposed that similar plants might be found in North America. After 3 months of laborious searching, Father Lafitau by accident did find plants that were similar to what Father Jartoux had described. The color of the fruit of the plant attracted the attention of Father Lafitau, and this was responsible for the discovery of American ginseng near Montreal in 1716.[7]

In 1718, Sarrasin published in the *Memoirs* of the French Academy an account of American ginseng. Father Lafitau also reported his discovery of ginseng plants in the same year. Soon the big news reached Peking. Father Jartoux was very much interested in this big discovery and came to Canada to see the differences between American ginseng and its Chinese counterpart. Some ginseng roots were gathered and sent as samples to China for examination. Acknowledgment was slow in coming, but finally, the word arrived: "our specimen lot is, indeed, ginseng; the quality is satisfactory." With this promise, soon the French began collecting ginseng, through Indians, for exporting. Soon the demand for ginseng thus created was so great that it became an important article of business in Montreal commerce.[8]

Before long, the short supply of wild ginseng in Canada became a serious problem. The ginseng trappers and traders spread the news of the ginseng business to the American colonies. The American wildlife hunters thus became enthusiastic about not only furs but also ginseng roots. About 30 years after the discovery of ginseng in Canada, American ginseng plants were found in many parts of the northern American colonies. Ginseng was found in southern New England in 1750 and in central New York and Massachusetts in 1751. It was similarly found plentiful in Vermont at the time of the settlement of the state.

Forest ginseng plants were eventually discovered in almost all states east of the Mississippi except in Florida, and in a few states west of the Mississippi River. It was growing wild in some parts of Maine, Connecticut, Rhode Island, Delaware, Pennsylvania, New Jersey, Maryland, Ohio, Virginia, West Virginia, Indiana, Illinois, Michigan, Iowa, Wisconsin, Minnesota, Kentucky, Tennessee, North and South Carolina, Georgia, Arkansas, and Alabama. Most of the wild ginseng was found in those states touched by the Allegheny Mountains.

The American ginseng plant is a small, unassuming perennial herb, 10–20 inches high. It is of very slow growth, even under the most favorable conditions. The plant is propagated from seeds, with stems bearing a single whorl of three palmately compound leaves, a solitary stalked umbel of greenish-white flowers, and bright red fruit. When the ginseng plant is old enough to produce fruit, it is rather noticeable and is easily recognized, but until 3 years old, it is not usually very prominent.

The seedlings at first somewhat resemble newly sprouted beans, in that they develop two cotyledons, and from between them a stem with two minute leaves. These enlarge until the plant has attained its first season's growth (about 2 inches). The growth of the plant during the first year is to develop the bud at the crown of the root, which is to produce the next season's stem and leaves. In autumn the stem dies and breaks off, leaving a scar, at the side of which is the solitary bud. In the spring of the second year, the bud produces a straight, erect stem, at the top of which one to three branch-like stalks of the compound leaves appear. Three to eight leaflets are developed, which usually rise not more than 4 inches from the ground. The third year, 8–15 leaflets may be put forth, and the plant may attain a height of 8 inches. In the succeeding years, the plant may produce three and five leafstalks 3–4 inches long, each bearing five thin leaflets (Figure 3.1).

Leaves are palmately arranged; two of them are smaller (1–2 inches long), the remainder are larger (3–4 inches), egg-shaped in outline, with the broad end away from the stem, abruptly pointed and raw-toothed. The leaves are bright green in summer, turning to yellow in the fall. The 5- to 6-year-old plant grows from 10 to 15 inches in height from the ground.[1]

At the point where the leafstalks meet, the main axis is continued into an erect but slender flower stalk, 2–5 inches long, bearing in early July or in late June a number of inconspicuous greenish-white flowers. These are soon followed by the fruit, which develops rapidly, remaining green until the middle of or late August, when it begins to turn sharp red, becoming scarlet and ripe in late September. The berries, which have the taste of the root, are the size and shape of small wax beans, and contain two or occasionally three seeds each.

No seed is produced in the first year, and only occasionally are berries found in the second year, and then only on extra strong plants in the garden. It is only in the third year that the plant produces seed in any quantity. Plants in cultivated beds produce more freely than those in the forest. The largest stalk of wild ginseng seldom produces more than 20 or 30 seed berries, and only a few of these can be expected to survive and germinate, since many birds, mice, and chipmunks are very fond of ginseng seeds. The seeds must never be allowed to become completely dry or they may fail to germinate. In the forest, however, the seeds ripen in the autumn, fall to the ground, and are covered for 18 months by the decayed leaves of the forest before the young seedlings appear. The seedlings must be properly protected and transplanted at least once before being set into the permanent beds. At all times, the beds must be kept free from weeds.

American ginseng usually has a thick, spindle-shaped root, 2–3 inches long or more, and about one-half to one inch in thickness, often branched, the outside prominently marked with circles or winkles. The root is simple at first, but after the second year it usually becomes forked or branched, and it is the branched root,

American Ginseng

Branch, root, flower, berries, and seeds of American ginseng.

FIGURE 3.1 American ginseng. (Adapted from Hou, J. P. *The Myth and Truth About Ginseng*, Cranbury, NJ: AS Barnes and Co., Inc., 1978.)

especially, that vaguely resembles the arms and legs of a human body that used to be of particular favor in the eyes of the Chinese buyers. But this irrational belief no longer holds true. Ginseng has a thick, pale-yellowish white or brownish-yellow bark, prominently marked with transverse wrinkles, the whole root fleshy and somewhat flexible. If properly dried, it is solid and firm. It has a slight aromatic odor, and the taste is slightly bitter followed by sweet and mucilageous. The dry root is about one-third the size of a fresh one. The best good-sized roots you find in the forest are from a few years to a few decades old. Though the roots acquire value from age, they do not increase much in size, 5 ounces (about 150 grams) being a giant root, but usually they are 1–2 ounces after dried. The dried root tastes slightly bitter and somewhat sweet afterward.[9]

The grading of ginseng roots is a highly subjective practice. The criteria are sources (wild or cultivated), shape, size, color (outside and inside), taste, texture, and markings. The trend has been toward continuously increasing in price since the beginning of the ginseng trade. A history of trade of American ginseng is given in the next chapter.

AMERICAN GINSENG ADVENTURE

In the state of nature, ginseng only propagates by seeds. If the plant is dug up prior to the ripening of the seed in September, it is deprived of its only means of perpetuating itself. Yet this was the very thing that happened to American ginseng. The *sang* diggers, a class of people who eked out a livelihood by hunting *sang* by shooting and trapping, exercised no judgement as to the season of digging. The plant was dug as soon as it was found in the forest, whether in April, May, or October. The other cause of the decreased supply of the wild root was the clearing of forest lands and the bruising and trampling of stock pastured in the woods. Both in Canada and in the United States, wild ginseng became extremely rare by the middle of the nineteenth century owing to extensive collecting without replacing. Nowadays the wild root is reported to have completely disappeared in many states where it used to be abundant. The ginseng plant seems to be extinguished entirely in Canada.

The visible decrease in the supply of the wild root and the constant increase in the demand as well as the increase in prices have led to many attempts to cultivate American ginseng by the *sang* diggers as early as in the early nineteenth century. But at that time failure was so frequent that its culture had once been declared impossible. However, such was not the case, since with proper attention and culture in an environment suited to its peculiarities, it may be grown successfully.

Until 1870, American ginseng was first successfully cultivated by Abraham Whisman at a place then called Boones Path in Virginia. Later George Stanton of Summit Station, New York, after long years of struggle, succeeded to grow the plant in the 1880s. In the beginning he transplanted a few small white seedlings and roots into his garden, and later he grew plants from the seeds of his own production. Not too long afterward, he became the first industrialized ginseng grower in the United States. He earned the title of "The Father of the American Ginseng Industry." In his 80-square-foot "George Stanton's Chinese Ginseng Farm" in 5 years, he produced 320 pounds of ginseng root, which when dried, would have been about 106 pounds, harvested. The production was sold for $575, which was a sizable amount of income out of such a tiny garden. He was so devoted to his ginseng cultivation that he treated his plants as "babies," as he called them.

It is more fascinating that A.R. Harding of Columbus, Ohio, began his cultivation of American ginseng in 1899. He gave up his regular medical practice and concentrated on ginseng experimentation. He had a wide understanding of cultivation, propagation, and marketing of this valuable root in addition to its uses in the treatment of his patients. He transplanted and cultivated ginseng plants from each state and tested the extract in many types of patients. He believed ginseng was useful for many illnesses wherein regular treatment had failed.

Between 1870 and 1895, there were about 20 ginseng gardens started. Ginseng farms mushroomed all over the Eastern and Midwestern states during the "ginseng boom period" between 1880 and 1903. However, a serious ginseng disease, Alternaria blight, broke out in 1904, causing great damage to the crops, and hundreds of young ginseng farms were terminated. In the United States, only about 600 acres of ginseng are planted. Ginseng farms in Georgia and Pennsylvania, but mainly in Wisconsin, are the principal cultivators. The average per-acre yield is about 1,000 pounds, and the total yearly harvest is from 100,000 to 150,000 pounds in the United States. The largest ginseng farm, that of the Fromm brothers of Hamburg, in western Wausau, Wisconsin, produces the most cultivated ginseng in the United States for domestic uses and exporting.

REFERENCES

1. Hart, B. L., in *Cyclopedia of Farm Crops*, L. H. Bailey, Ed. Macmillan, New York, 1922.
2. Fournier, P., *La Nature*, (Paris), 1940, p. 244.
3. Harding, A. R., *Ginseng and Other Medicinal Plants*, Emporium, Boston, 1972.
4. Baugart, R. A., *Milwaukee Journal*, December 13, 1974.
5. Schroger, A. W., "Ginseng: A Pioneer Resource," *Wisconsin Aca. Sci.*, Arts and Lett., 57: 65, 1963.
6. Garriques, S. S. *Ann Chem. Pharm.*, 90: 231, 1854.
7. Vogel, V. J., *American Indian Medicine*, University of Oklahoma Press, Norman, 1970, pp. 307–310.
8. Stockberger, W. W., "Ginseng Culture," in *Farmers Bulletin No. 1184*, U.S. Department of Agriculture, Washington, D.C., U.S. Government Publishing Office, 1921, revised 1949.
9. Williams, L., "Growing Ginseng," in *Farmers Bulletin No. 2201*, U.S. Department of Agriculture, U.S. Government Printing Office, 1964, revised 1973.

4 Ginseng in Korea

HIGHLIGHTS OF KOREAN GINSENG

Korean ginseng is botanically identical to the ginseng plant growing in Manchuria and in the maritime province of Siberia. Historically, Korean ginseng plants grew abundantly in the forested mountains over the peninsula. Wild mountain ginseng is called *san-sam*. Korean ginseng has been a very valuable remedy in medicine since the sixth century. In modern times, Korea has become the major ginseng-producing country in the world. This may be ascribed to the fact that Korea has the longest history in producing ginseng. Moreover, the climate and geographical conditions may be optimal for ginseng cultivation. Korea deserves the title of the "ginseng country."

According to legend, Korean history began some 4,000 years ago. The majority of the Korean people were immigrants from North China and Manchuria. About AD 100 the kingdom of Koguryo, the first truly Korean state, emerged in the middle Yalu region. During the fourth century, the tribes of the southern half of the peninsula had coalesced into three federations, known as the Three Han. Paekche in the southwestern or Ma Han area, Silla in the southeastern or Pyon Han region, and the Chin Han region, which lays across the center of the peninsula, fell to the hands of Koguryo in the early fifth century. In the ensuing struggle among the Three Kingdoms of Koguryo, Paekche, and Silla, it was Silla that won control of the peninsula under the help of the Chinese army during T'ang dynasty. Then came the Silla Unification Period. During the Yi dynasty (1392–1910), King T'aejo adopted the name of "Chosen for Korea," and moved the capitol from Kaesong to Seoul. The dynasty he established lasted 500 years until the annexation of Korea by Japan.[1]

As a result of the historically close association between Korea and China, Chinese medicine was prevalent in Korea, and Korean ginseng was known to the Chinese as early as in the third century (during the Wei dynasty). Even in the very beginning, the Koguryo kingdom envoys made 92 diplomatic trips to China, and each time Koguryo ginseng was brought to the Chinese emperor as the most valuable gift. In the sixth (Sui dynasty) and seventh to ninth centuries (T'ang dynasty), ginseng roots from Paekche and Silla were similarly offered to the Chinese emperors on numerous occasions. Ginseng root was used by emperors and high officials in ancient China as the most valuable medicine.

According to Chinese medical literature, the three different types of ginseng (root) from the Three Kingdoms are not all equal in appearance, color, taste, and activity. The great Chinese physician Tao Hung-ching recorded the descriptions of the ginsengs in his book as follows: "The ginseng (root) from Paekche is slender, round in shape, firm and with lighter taste than that of *tang-shen* (the genuine ancient Chinese ginseng); ginseng from Koguryo is large, soft, and inferior to *tang-shen* and that from Paekche. The ginseng from Silla is yellowish with light taste, and more closely resembles the human figure."[2]

CULTIVATION OF KOREAN GINSENG

Growing ginseng plants was first started in the Ching-shan and Chuan-lo provinces in South Korea in the sixteenth century. Later ginseng was cultivated on large scales in Kei-jo (Kae-song) and Kum-san. New ginseng cultivation centers are Kang-gye, Kum-kand, Kyan-ji, Chung-chu, and Che-ju. In the beginning of the twentieth century, the production of Korean ginseng reached 700,000 pounds per year, valued at more than 2 million yen. Most of this was exported to China, Japan, and other southeastern Asian countries (Figure 4.1). Even at that time, the top-grade Korean

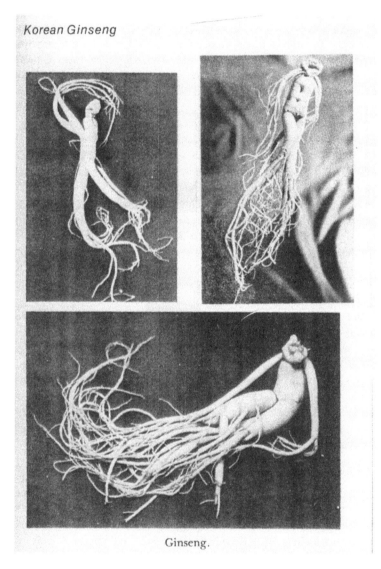

Korean Ginseng

Ginseng.

FIGURE 4.1   Ginseng. (Adapted from Hou, J. P. *The Myth and Truth About Ginseng*, Cranbury, NJ: AS Barnes and Co., Inc., 1978.)

red ginseng root was sold at about \$30–\$35 per *catty,* which was about seven times higher than the price of imported American ginseng, or about twice that for Korean white ginseng.[3]

Unfortunately, a disastrous effect of the ginseng disease that broke out in 1902 in Korea crippled the business for about 11 years and threatened to extinguish the entire trade, and there were no exports until 1910.[4]

The climate and soil of the Korean Peninsula are peculiarly suited for ginseng cultivation. More and more farmers attempted to grow ginseng plants for higher profit. At present there are 33 ginseng-growing districts in South Korea including Kang-wha, Kim-po, Kum-san, and Pu-yo, whose annual production of crude ginseng amounts to 5,700,000 kg. This figure was expected to increase to 15 million kilograms in 1981 according to the Korean Office of Monopoly. The total cultivation area of red ginseng products is about 16 million square meters. With such industrialized cultivation, a supply shortage in ginseng is not expected in the next 10 years in Korea's ability to support the world market.[5]

GINSENG LAW

Korea has a long history in governing and controlling ginseng production, particularly the red ginseng root. The control of the production and sale of red ginseng was initiated in the seventeenth century (the Yi dynasty, AD 1606). Probably the first government regulation of ginseng was promulgated in 1686 and the second in 1707. From 1897 to 1910, red ginseng production and trade were controlled directly by the emperor's office. In 1868, taxation on ginseng cultivation in the Gae Seung area began.

A "modern monopoly law" and its bylaws for red ginseng were established in 1907. There were revisions in these laws in 1920, and they were effective until 1945. The major regulations of the ginseng monopoly law included registration of seeds, and intensifying the penalty for illegal production of red ginseng. During the Japanese annex period (1910–1945), the Office of Monopoly, located in Kaesong, was controlled by the Japanese (Figure 4.2).

On August 15, 1947, Korea was divided into South and North Korea. The famous ginseng city Kaesong was divided into two parts. As a result of social instability, ginseng production was suspended for quite a number of years. Fortunately, Korean ginseng production has been revitalized since 1960. In order to promote the quality of ginseng and its products uniformly, the monopoly law of red ginseng of December 1959 was radically revised in December 1972. A new regulation on white ginseng and its products was simultaneously promulgated. The production and sale of white ginseng products then became privately owned operations that were also subject to control by the 1972 ginseng law.[3]

REMEDIAL USE OF GINSENG IN KOREA

Although ginseng has been recognized by the Koreans as a useful remedy dating back to the Three Kingdoms period, it became a recognized medicine only in the Yi dynasty. Ginseng was used in Korea, as in China, for its alternative, tonic,

FIGURE 4.2   Ginseng plant with fruits. (Adapted from Hou, J. P. *The Myth and Truth About Ginseng*, Cranbury, NJ: AS Barnes and Co., Inc., 1978.)

stimulation, carminative, and demulcent properties. It has reputed virtues for all forms of debility and dyspepsia, spermatorrhea, persistent vomiting of pregnancy, old cough, and polyuria, which are treated with ginseng preparations for relief and cure.[4]

In the early tenth century, ginseng was also used in making food and cosmetic products. Ginseng drinks, tea, candy, and cosmetics preparations, such as toothpaste, bath oil, and creams, were the most commonly processed ginseng preparations.

Ginseng's tonic effect has also been tested in the veterinary field. It was said that in the sport of horse racing, ginseng was given in the horses' feed for several weeks before racing. The horses under ginseng treatment showed better performance than those without.[6]

MODERN KOREAN GINSENG PRODUCTS

Ginseng root as it is dug up is called "watery ginseng" (*su-sam*) from which white ginseng root (*pae-sam*), yellow ginseng root (*hwang-sam*), and red ginseng root (*hong-sam*) are processed. White ginseng root is made by initially peeling the watery root with a bamboo knife after it has been cleaned, and drying it in the sun, or when the weather is unfavorable, drying over a charcoal fire or in an oven (Figure 4.3). Yellow ginseng is made by steaming the peeled watery root and then drying it. This type of product is less desirable. The making of red ginseng root includes steaming the watery root for about 3–4 hours, dehydrating it in a dry room, and parching it in the sun. Red ginseng is monopolized by the Korean government, and cultivators are prohibited from making and selling it. White ginseng looks slightly yellowish-white, while red ginseng is somewhat translucent and reddish-brown in color. Both red and white ginseng are packaged in sealed cartons marked with name, grade, total weight, number of roots, and source of production printed on the box.[7]

Modern ginseng products are being developed to meet the government standards for exportation. The newer products are honey ginseng, ginseng extract, ginseng powder, ginseng pills, ginseng capsules, ginseng tablets, ginseng tea, ginseng syrup, ginseng cakes and candies, ginseng tincture, ginseng elixir, ginseng fluid extract,

FIGURE 4.3 Sun-dried raw ginseng. (Adapted from Hou, J. P. *The Myth and Truth About Ginseng*, Cranbury, NJ: AS Barnes and Co., Inc., 1978.)

and ginseng alcoholic beverages.[8] The most popular ginseng tonic preparations, such as ginseng in wine (*In-sam-Ju*), ginseng chicken essence, ginseng concentrate, and ginseng-vitamin-mineral combination preparations, are presently the most desirable marketed products.

White Ginseng Root

This is one of the most popular ginseng products of South Korea to date. It is packaged into 600-, 300-, 150-, and 75-g boxes. The largest roots weigh about 47 g per piece, the medium-sized roots weigh about 10 g per piece, and the smallest roots weigh only 3 g per piece. As a result of the sizes, the price of the white ginseng varies significantly.

Honey Ginseng

Honey ginseng is white ginseng root preserved in syrup. It gives a palatable taste.

Ginseng Tea

Instantly soluble ginseng tea granules are packaged in foil bags or in bottles. The granules contain about 8% of dry ginseng extract and soluble inert materials such as lactose and soluble starch. It gives a stimulant effect with typical ginseng flavor and fragrance.

Ginseng Extract

Ginseng extract is the most potent ginseng product for tonic effect. It contains about 40% water and 60% solid materials including extracted principles from the root.

Ginseng Capsules

Ginseng capsules contain dried ginseng extract and inert fillers.

Cosmetics and Animal Health Products

In the cosmetic area, creams and beauty lotions have been marketed for rejuvenation and a youthful-looking effect. In the field of animal health, ginseng can also be used in many cases for an antifatigue effect, and homeostasis, and may produce longevity as well.

REFERENCES

1. Wagner, E. W. in *The Encyclopedia Americana*, Vol. 16, Americano Corp., New York, p. 528 f.
2. Li, S.-C., *Pen-ts'ao Kang-mu*, Shang-wu Press, Shanghai, Vol. 12 (original), 1970.

3. Yoon, J. H. in *"Korean Ginseng Science Symposium,"* Korean Society of Pharmacognosy, Seoul, 1974, pp. 227–239.
4. Read, B. E., *Pharm. J.*, 117: 671, 1926.
5. *"Koreanishe Ginseng,"* a Foreign Press article of Korean ginseng, Office of Monopoly, Seoul, 1974.
6. Imamura, T., *Nin-jin Shin-so*, Office of Monopoly, Kae-jo, Korea, 1923.
7. "Korean Ginseng," in *Abstract of Papers of First Ginseng Symposium*, Seoul, Korea, September 1974.
8. Kim, S. K. in *"Korean Ginseng Science Symposium,"* Korean Society of Pharmacognosy, Seoul, 1974, pp. 207–226.

5 Ginseng in Japan, Siberia, and *Panax Notoginseng*

GINSENG IN JAPAN AND CHIKUSETSU NIN-JIN

The beginning of the Yamato era in the third century marked the unification of ancient Japan. At the time, Japan was a young, struggling country, while its neighbor, China, had been enjoying a highly developed culture and social life. The Yamato government sent the first envoy to China in 607 and thus established a diplomatic relationship with the emperor of the Sue dynasty. Later in the T'ang dynasty, the Japanese government dispatched students, monks, and professionals as well as envoys to China to observe and learn Chinese culture and technology. After the Taike reform in 645, Japan adopted the Chinese written language, arts, crafts, social and government systems, and medicine. In 710, the first permanent capitol, Heijo (later called Nara), was built after T'ang's capitol of Ch'ang-an. In the Japanese society, it became stylish to copy the gorgeous silk robes worn by the Chinese nobility, and to imitate their customs and social manners.[1] Chinese medicine was adopted and practiced in Japan, known as Kanpo-I in Japanese. Chinese medicine occupies a special place in Japan even today.

Nin-jin means ginseng, while *Panax ginseng* is called *Otane-ninjin* in Japanese. Though ginseng has been imported to Japan from Korea (the Kingdom of Paekche) and China since the seventh century, ginseng started to be used as a medicine in the eighth century. At that time, ginseng was the most valuable drug, and it was controlled by a special office of the emperor. The price of ginseng was extremely high, and it was available only to the palace, the nobles, and the rich. Ginseng became more popular after the ninth century due to more Korean ginseng being available. Chinese medicine in Japan became the flourishing medical practice after the ninth century, and lasted for about 1,000 years until the Meiji restoration in 1868.

In early Japanese medical practice, ginseng was used for tonic, stomachic, many forms of debility, cardiac stimulant, thirst, diabetes, vomiting, coolness of limbs, etc. It was used alone or with other drugs as in the popular prescriptions in Chinese medicine. Ginseng was used in many special preparations, such as cosmetics, drinks,[2] tea, and candy, and in preparations for aphrodisiac properties with deer's antlers, cinnamon, seal's kidney, and musk as described in the Japanese books *Nin-Jin -Shih* (*History of Ginseng*).

There is no evidence that Japan had wild mountain ginseng in ancient times. Japan imported ginseng seeds from Korea in 1607, and later a large amount of seeds, seedlings, and even live roots were imported from Korea and China and planted in quite a number of counties in Japan where the weather, soil, and terrain were suitable. Hokkaido, Yamagata, Nagano, Shimane, and Fukushima were the early ginseng cultivation centers. Ginseng was called "Imperial ginseng," and after a few failures, ginseng farming finally succeeded. In 1907, ginseng cultivation became a good business, and thousands of farmers grew ginseng, which was distributed in 43 counties. However, this boom did not last too long as a ginseng epidemic occurred, and most of the farms then died away.[3]

In modern times, ginseng in Japan is cultivated in Fukushima, Nagano, and Shimane counties. In the beginning, the annual production of ginseng in Japan was about 70,000–200,000 kg. The latest Fuji marketing report shows that growing ginseng is still a fairly profitable business. Ginseng farmers are not growing ginseng exclusively but are regular rice growers. From 1969 to 1975, ginseng farming areas were increased from 383 hectares in 1969 to 413 hectares in 1972 to 425 hectares in 1975, although the number of farms actually decreased from 4,590 in 1969 to 4,040 in 1972. Most of the ginseng farms are small, individual gardens, and about 80%of the total 4,000–5,000 farms are located in Nagano county. Collective cultivation of ginseng was only recently conducted in Shimane county.[4]

The actual ginseng cultivation areas in the three counties in 1973 were 82 hectares (19%) in Fukushima, 250 hectares (58.5%) in Nagano, and 96 hectares (22.5%) in Shimane county. (One hectare is 2.47 acres.) The total ginseng farming area in Japan is about 1,057 acres.

Total annual production is about 319,000 kg or 703,267 lbs divided into 54,000 kg from Fukushima; 185,000 kg from Nagano; and 80,000 kg from Shimane county. Accordingly, the production of ginseng in Nagano alone is about 58% of the total ginseng produced in Japan.

The fresh ginseng roots are processed to make commercially useful products. The majority of the roots are made into red ginseng according to the classical Chinese methods and are then exported to China and Hong Kong. The rest of them are made into white ginseng and other processed products for domestic consumption and exporting.

In the early twentieth century, Japan had no problem getting Chinese and Korean ginseng, for Japan annexed Korea to its protectorate in 1907, and later occupied Manchuria in 1923 until the end of World War II in 1945. During the Japanese military occupation of the ginseng country and ginseng-rich Manchuria, the entire ginseng business was monopolized by the Japanese government. Also during the occupation years, Japanese scientists seriously studied ginseng.

Chikusetsu nin-jin, or Japanese ginseng, has been native to the mountain areas of Japan since ancient times. The word *chikusetsu* means that the ginseng root resembles bamboo knots. Botanically, it is *Panax japonicum* C.A. Meyer, of the Araliaceae family. It has also been called To-jin, meaning *native ginseng*, and *tochiba ninjin*, in Japanese.

The part of the Japanese ginseng plant that is above ground is similar to *Panax ginseng*, but the root is not. Japanese ginseng root tastes more bitter than ginseng, and it is much cheaper. For thousands of years, Japanese ginseng has been used as a ginseng substitute for an expectorant, stomachic, and antipyretic agent in the Orient.

GINSENG IN SIBERIA

Panax ginseng also is found naturally in the maritime areas of Siberia. In ancient times, ginseng grew over vast territories. At present, it grows in considerable amounts only in the territories of the Soviet Union, where state laws prevent it from extermination. A special state farm called *Zhenshen* was founded to grow this most valuable medicinal plant. Scientific establishments and individual ginseng growers in the Caucasus, and the Ukraine, in the Baltic republic, and in Siberia have also amassed a vast experience in growing ginseng.[5]

Eleuthero, botanically known as *Eleutherococcus senticosus* Maxim of the Araliaceae family, is called *Siberian ginseng* under the popular name.

E. senticosus has a history of use in traditional Chinese medicine where it is known as *cì wǔ jiā*. In regard to the species' scientific name, *Eleutherococcus* (from Greek) means "free-berried," and *senticosus*, from the Latin word *sentis* (thornbush, briar) is an adjective meaning "thorny" or "full of briars or thorns." Because its branches are spiked with thorns, it has nicknames of *touch-me-not* and *devil's bush*. Siberian ginseng is not known in Oriental medicine, although it is abundant in eastern Siberia, in Korea, and even in the provinces of Shan-si and Ho-pei of northern China. But the naturally growing area of *Eleuthero* in Siberia exceeds 10 million hectares.[6]

It is a shrub about 2 m in height with numerous thin thorns. The leaves are long stalked, palmately parted, and similar to those of ginseng. The flowers are small (female is yellow, male is violet in color), gathered, globular, and umbrella shaped. It blooms in July and bears oval fruit; the fruit turns black when ripened in September.

Scientific studies on *Eleuthero* in the Soviet Union found that Siberian ginseng is of negligible toxicity and has anabolic, gonadotropic, stimulating, protective, and adaptogenic actions. One of the numerous useful effects of the Siberian ginseng preparations is the antistress action.[7] In this regard, Siberian ginseng is similar to *Panax ginseng* (Figure 5.1).

As Siberian ginseng takes only a year to grow, it is available in larger quantity than the Panax ginseng in the market. Siberian ginsengs are available year-round. Moreover, Siberian ginseng is cheaper than Panax ginseng.

The two are also different in their active ingredients. Panax ginseng contains the active ingredient ginsenosides, while Siberian ginseng does not contain ginsenosides. Instead of ginsenosides, Siberian ginseng contains the active ingredient eleutherosides. While Panax ginseng has warm, bitter, pungent characteristics, Siberian ginseng has a warm and mild character.

Siberian ginsengs are classified to the group of adaptogens, which raise resistance to various negative factors: physical, chemical, biological, and psychological. The preparations of adaptogens stimulate physical and mental ability and raise an organism's resistance to various kinds of sicknesses, poisoning, and irradiation. Although a relatively new addition to Western natural medicine, Siberian ginseng has quickly gained a reputation similar to that of the better known and more expensive Chinese ginseng. Unlike many herbs with a medicinal use, it is more useful for maintaining good health rather than treating ill health. Siberian ginseng produces a comprehensive strengthening and toning impact; it has been recommended in treating various neural diseases, impotence, lung ailments, medium forms of diabetes mellitus, and malignant tumors. They stimulate the central nervous system, stimulate sex gland activities, decrease sugar and cholesterol levels in blood, improve appetite, and sharpen sight and hearing.

Research has shown that it stimulates resistance to stress, and so it is now widely used as a tonic in times of stress and pressure. Regular use is said to restore vigor,

FIGURE 5.1   Siberian ginseng. (Adapted from Hou, J. P. *The Myth and Truth About Ginseng*, Cranbury, NJ: AS Barnes and Co., Inc., 1978.)

improve memory, and increase longevity. It has been used during convalescence and in the treatment of menopausal problems, geriatric debility, physical and mental stress, and a wealth of other ailments.

Siberian ginseng or eleuthero has been used in China for 2,000 years as a folk remedy for bronchitis, heart ailments, and rheumatism, and as a tonic to restore vigor, improve general health, restore memory, promote healthy appetite, and increase stamina. Referred to as *ci su ju* in Chinese medicine, it was used to prevent respiratory tract infections as well as colds and flu. It was also believed to provide energy and vitality. In Russia, eleuthero was originally used by people in the Taiga region of Siberia to increase performance and quality of life and to decrease infections.

The ability of Siberian ginseng to increase stamina and endurance led Soviet Olympic athletes to use it to enhance their training. Explorers, divers, sailors, and miners used eleuthero to prevent stress-related illness. After the Chernobyl accident, many Siberian citizens were given eleuthero to counteract the effects of radiation.

PANAX NOTOGINSENG

Panax notoginseng is a species of the genus *Panax*, and it is most commonly referred to in English as *notoginseng*. In Chinese it is called *tiánqī, tienchi ginseng, sānqī* or *sanchi*, three-seven root, and mountain paint. Notoginseng belongs to the same scientific genus as Asian ginseng. In Latin, the word *panax* means "cure-all," and the family of ginseng plants is one of the best-known herbs.

Notoginseng grows naturally in China and Japan. The herb is a perennial with dark-green leaves branching from a stem with a red cluster of berries in the middle. It is both cultivated and gathered from wild forests, with wild plants being the most valuable. The Chinese refer to it as "three-seven root" because the plant has three petioles with seven leaflets each. It is also said that the root should be harvested between 3 and 7 years after planting it.

It is classified in Chinese medicine as warm in nature, sweet and slightly bitter in taste, and nontoxic. The dose in decoction for clinical use is 5–10 g. It can be ground to powder for swallowing directly or taken mixed with water: the dose in that case is usually 1–3 g.[8] In the *Bencao Gangmu* (*Compendium of Materia Medica*, AD 1596), it is stated: "On account of the fact that sanqi is a herb belonging to the blood phase of the yang ming and jue yin meridians, it can treat all diseases of the blood." Notoginseng is an herb that has been used in China quite extensively since the end of the nineteenth century.[9] It has acquired a very favorable reputation for treatment of blood disorders, including blood stasis, bleeding, and blood deficiency. It is the largest ingredient in (*Yunnan Bai Yao*), a famous hemostatic proprietary herbal remedy that was notably carried by the Viet Cong to deal with wounds during the Vietnam War.

The plant is named for the irregular growth of its leaves. It was said that there were three leaflets on the right and four leaflets on the left of the stem, and they grew directly out of the apex of the stem. But this is not the description of *Ren-sen-san-ch'i* today.[10]

This plant grows naturally and is cultivated in the southwestern part of China. The primary area of natural production is in the forest mountains of the southeastern Yun-nan province and its adjacent areas of Kwang-si province, on the slopes of 1,000–2,000 m, particularly in Wen-shan, Kwang-nan, His-chow, and Yen-shan in Yun-nan province and Tien-gang and Chig-hsi in Kwang-si province (Figures 5.2 and 5.3). About 90% of the commercial product of *San-ch'I* ginseng today is from Yun-nan province. For this reason, the crude drug has been called *Yun-nan San-ch'i*. However, the *San-chi* has also been cultivated on slopes at altitudes of 1,000 m and above in Kwan-si and Yun-nan provinces. Loose, acidic soil, rich in humus is preferred. The natural plant of (*notoginseng*) ginseng can also be found in Sze-chwan, King-si provinces of China, North Vietnam, the Himalayas, and northern India.

FIGURE 5.2 Tienchi ginseng (*Panax notoginseng*). (Adapted from Hou, J. P. *The Myth and Truth About Ginseng*, Cranbury, NJ: AS Barnes and Co., Inc., 1978.)

The growth habit of *Ren-shen-San-ch'I* is similar to that of ginseng. The crude drug is the properly dried root of a 4- to 6-year-old plant. The root of *San-ch'I* ginseng is fleshy, firm, round, or conical in shape, with a smooth skin, brownish-gray in color, and the root is about 2–4 cm long and 1–2 cm in diameter. It tastes bitter first and then slightly sweet.[10]

The root of *P. notoginseng* contains 12% of saponins. It has the same number of ginsenosides, such as ginsenosides Rb_1–Rb_3, Rc, Rd, Re, Rg_1, Rg_2, Rh_1, F_2, and glucoginsenoside Rf as in ginseng, the root of *P. ginseng*.

In addition, some new saponins named *notoginsenosides*[11] were isolated from the root. They are notoginsenosides R_1, R_2, R_3, R_4, R_6, and gypenoside XVII.[12,13]

FIGURE 5.3 Radix notoginseng. (Adapted from Hou, J. P. *The Myth and Truth About Ginseng*, Cranbury, NJ: AS Barnes and Co., Inc., 1978.)

Notoginsenoside R_1 is the major notoginsenoside of the herb and is used to differentiate from the ginseng root in the *Chinese Pharmacopoeia*. From the rootlet of *P. notoginseng*, sanchinosides B_1 and B_2 were isolated.[14] The leaf, flower, and seed of the plant also contain ginsenosides and notoginsenosides.[15,16]

The root of *P. notoginseng* contains essential oil, and the major components of the oil have been identified as α-guaiene, β-quaiene, and octadecane.[17]

Pharmacologically, *Ren-shen-san-ch'i* is a very effective agent in arresting hemorrhage and bleeding in wounds, including snake and tiger bites. Internally, it has been prescribed in hematemesis, menorrhagia, etc. The leaves also have similar properties and are often combined with the root for medicinal preparations.

REFERENCES

1. Hall, J. W. in *The Encyclopedia Americana*, Vol. 15, Americano Corp., New York, pp. 814–824.
2. Imamura, T., "Medical and Pharmaceutical Aspect," in *Nin-jin shih, (History of Ginseng)*, Vol. 5, Shibunkaku, Kyoto, 1971.
3. Imamura, T., *Ninijin Shin-so*, Office of Monopoly Press, Kae-jo, Korea, 1923.
4. *Fujisan Marketing Report*, July, 1975.
5. Brekhman, I. I., *Ind. J. Publ. Health*, 9: 148, 1965.
6. Sandberg, F., *Plant. Med.*, 24: 392, 1974.
7. Brekhman, I. I. in *Pharmacology of Oriental Plant*, K. K. Chen and B. Murkerji, Eds., Pergamon Press, New York, 1965.
8. Bensky, D., S. Clavey, E. Stoger, and A. Gable, comps. and trans., *Chinese Herbal Medicine: Materia Medica*, 3rd ed., Eastland Press, Seattle, WA, September 2004.

9. Dharmananda, S., "Rare Reactions to a Safe Herb: *Sanqi* (*Panax notoginseng*)." ITM Online (http://www.itmonline.org/arts/sanqi.htm).

10. Hsu, K. J., Dhou-hdun, *Pharmacognosy*, Hong Kong Wei-sheng Press, Hong Kong, 1971, pp. 458–459.

11. Tang, W. and G. Eisenbrand, "Chinese Drugs of Plant Origin," in *Chemistry, Pharmacology, and Use in Traditional and Modern Medicine*, Springer-Verlag, Berlin, Heidelberg, 1992, pp. 745–751.

12. Matsuura, H., R. Kasai, O. Tanaka, Y. Saruwatari, T. Fuwa, and J. Zhou, "Further studies on dammarane saponins of Sanchi-Ginseng," *Chem. Pharm. Bull.*, 31: 2281–2287, 1983.

13. Zhou, J., M. Z. Wu, S. Taniyama, H. Besso, O. Tanaka, Y. Saruwatari et al., "Dammarane-saponins of Sanchi-Ginseng, roots of *Panax notoginseng* (Burk.) F. H. Chen (Araliaceae): Structures of new saponins, notoginsenoides-R_1 and R_2 and identification of ginsenosides-Rg_2 and Rh_1," *Chem. Pharm. Bull.*, 29: 2844–2850, 1981.

14. Wei, J. X., L. A. Wang, H. Du, and R. Li, "Isolation and identification of sanchinoside B_1 and B_2 from rootlets of *Panax notoginseng* (Burk.) F. H. Chen," *Acta Pharm. Sin.*, 20: 288–293, 1985.

15. Taniyasu, S., O. Tanaka, T. Yang, and J. Zhou, "Dammarane saponins of flower buds of *Panax notoginseng* (sanchi Ginseng)," *Plant. Med.*, 44: 124–125, 1982.

16. Wei, J. X., L. Y. Chang, J. F. Wang, E. Friedrichs, M. Hores, and H. Puff, "Two new dammarane saponins from leaves of *Panax ginseng*," *Plant. Med.*, 45: 167–171, 1982.

17. Lu, Q. and X. G. Li, "Studies on the chemical constituents of the essential oil in the Renshn Sanqi (*Panax notoginseng*)," *Chinese Tradit. Herb. Drugs*, 19: 5–7, 1988.

Section II

The Flourishing Ginseng Business

6 The Flourishing American Ginseng Business in Chinese Market

AMERICAN GINSENG IN CHINESE MARKETS

Soon after the Chinese agreed to purchase American ginseng, a kind of ginseng stampede was ignited. Thousands of French Canadians and Indians took to the woods in the quest for easy-to-dig "amber nuggets." A ginseng trading company was formed almost overnight. The rush for "green gold" was on. In Quebec the root was purchased at 2 francs a pound by the fur traders who collected and exported it to China. In China, the same root was sold for as much as 25 francs. Ginseng trade with China was at that time controlled by the company of the Indies. In the beginning, the ginseng business was handled by the officers of the company. In 1751, seeing that the commerce in the root was so great, the company prohibited private venture on the part of its officers and assumed control itself. In only a short time, the price of ginseng was advanced from 12 francs to 33 francs a pound.

Demand became so great that the *sang* (American ginseng) hunt became a highly profitable venture, particularly for the French Canadians. Exportation of the roots quickly increased: often the shrewd tradesmen unscrupulously received an average of 10–12 times as much for *sang* as they paid uncomplaining diggers for it. The years rolled by, and the tons upon tons of *sang* were ripped from Canadian soil. When finally demands could scarcely be met, many *sang* shipments started to yield inferior roots—roots were harvested out of season and improperly dried, leaving them visibly scorched.

In 1752 Chinese buyers inspected several shipments of American ginseng from Canada and found that the roots had been improperly treated and that a good portion of the roots was *not even ginseng*. This unprincipled practice all but ruined the Canadian *sang* market. Chinese traders were furious over the attempt to deceive them, and ginseng exports, in a year, dropped from over $100,000 to only $6,500. The ginseng trade in Canada thus dwindled rapidly and quietly faded away.[1]

The spark of demand for American ginseng was to remain dim for nearly 30 years, but gradually a limited and exceedingly wary market started once again to open its doors, and the favorable focus was on the American colony's ginseng roots. Ginseng grew naturally throughout the forest of most northern states, particularly the Appalachian Mountain areas, and the Chinese traders had confidence in the colonial roots; a stimulating business followed. Early American settlers often collected *sang* as a sideline and for a quick cash return. Local merchants accepted it in exchange for supplies, although fur buyers were the principal dealers.

The gathering and marketing of the root began in a small way but picked up momentum when it was found that the range of the plant extended throughout the Eastern United States. Usually the dry ginseng roots were purchased by the fur dealers in New York as well as other large cities. These buyers either disposed of their holdings to Chinese agents or exported directly to Hong Kong, the principal port of American goods entering China. There the roots were purchased in large quantities by the traders to supply the retailers or drugstores.

For many years, New York, Boston, and Philadelphia served as the main collection points for exportation of *sang*. The initial export from the United States to China started in the mid-eighteenth century, and the earliest trades were through the East India Company in England. A shipload of 55 tons of ginseng sailed from Boston to China in 1773. In 1782 the first direct shipment of American ginseng to China was made by John Jacob Astor. The root from that shipment was reported to have brought $3 a pound. In February 22, 1784, another American ship, *Empress of China*—a 360-ton ex-privateer—made another direct sail to China. Ginseng was the principal cargo. Daniel Boone, the great American pioneer and frontiersman, also collected and dealt in ginseng. In 1787 he started up the Ohio River with a 15-ton boatload of ginseng from Philadelphia on its way to China. Unexpectedly, he ran into trouble—the boat overturned.[2] Later the Chinese herbalists on the West Coast became the main ginseng dealers and exporters. During the end of the eighteenth century and beginning of the nineteenth century, American ginseng for export was mostly collected from the states of Pennsylvania, Ohio, Kentucky, and New York. A Kentucky buyer of ginseng noted that he had bought 1,700 pounds and he wanted to buy 300 pounds more to make up a ton.[3] Such a large amount of dry root would bewilder a modern ginseng hunter and no doubt represented a collection acquired over an extended period, even then. In the 1790s, the wild root was selling for as high as $1 a pound and about 200,000 pounds were exported each year to China.

The cultivation of transplanted ginseng seemed to have started sometime in the 1840s. Due to the presence of the unnatural fore-stall ginseng in the shipment, the price was dropped from 68¢ in 1841 to 44¢ in 1842 and to 34¢ per pound in 1843. The Chinese buyers do not appreciate the cultivated species even though they are heavier and larger than the wild species. In 1858, about 366,053 pounds of ginseng, valued at $193,736, were exported to China. The price was 52¢ per pound. The export of *sang* continued through the years, and the price continued to rise.

When the population moved westward, ginseng was discovered in abundance in the territory immediately west of Mississippi. About 1845, Green County, Wisconsin, had acquired the reputation of being *sang* county.[4] Discovery of the American ginseng plant in many other counties followed, and by 1860 large amounts of Wisconsin wild ginseng were being collected and exported to China.[5] Later, ginseng collected in Minnesota was also exported to China.[6] The market price of ginseng in 1862 was then 80¢ and more a pound. By 1868 the market value of ginseng had doubled, and in that year, over 370,000 pounds of American ginseng reached Chinese buyers. The price was even higher in 1878, and by 1888 it once again had doubled to $2 and more a pound. But by now the inevitable was beginning to happen, and wild ginseng was becoming much more difficult to find. Even with the stimulation of higher prices, the year 1888 noted at least 112,000 pounds less than 1878; and in 1889 and 1890, with prices still higher, exported quantities took another sharp drop. In the year 1896, the price of wild ginseng climbed to $3.86 a pound, with only 199,000 pounds harvested. In 1897 only 179,573 pounds were exported, though the price jumped to the all-time high of $4.71, and the total value of ginseng exported to China reached $850,000. By the year 1898, the supply of wild ginseng became more scarce, and many ginseng farmers began to harvest their cultivated product. Such extensive digging in quantity without even considering the age and without replacement contributed to the near extinction of the wild roots in the late nineteenth century.[1]

As a result of the scarcity of the wild root, the price rose again to $5.20 a pound in 1900, and then $6.35 to $7.25 in 1907. The inflation was about 5%–10% per year. From that time on, more and more cultivated roots helped to fill the increasing demand for wild ginseng, which steadily became less and less available. Cultivated ginseng, tagged as such, was marketed, and the price dropped once again from $4.70 to only $3.66 a pound.[7] However, due to the infinite demand for ginseng, soon the price jumped up to $7.00 again in the early twentieth century. The amount exported kept pace with the price, and the total export of *sang* exceeded $1 million after 1905.

As a result of the attractive price, more and more people were involved. A ginseng boom took place in the United States in the years 1895 to 1904. Hundreds of new gardens were started, and stock companies were formed to grow and trade ginseng plants. Ginseng seeds were selling for about 8¢ each or $6 per ounce or $90 per pound.[6] Unfortunately, a leaf disease of ginseng became prevalent in 1904, plaguing many plantations and discouraging the inexperienced growers.

Even during the First World War, the price of ginseng did not drop. By 1922, the market value of wild *sang* reached $11.51 a pound, and top-grade large roots were bringing an even better price, while the cultivated roots were usually half the price of the wild species. The total exports valued more than $2 million after 1922. The price of American ginseng remained high during the 1920s.

From Figures 6.1 and 6.2, one can see that the demand for American ginseng has constantly increased since the trade started in the early eighteenth century, not only the export and total valuation but also the average price of American ginseng exported inflated from 1821 to 1975.[8]

During the Sino-Japanese War and the World War II years, there was little American ginseng exported to China, although the amount did reach 185,976 pounds in 1946; a downward trend appeared in 1947 and later years. But the demand for American ginseng began to increase after 1950, and also most of the ginseng was exported to Hong Kong instead of to China. In 1951, the market value of American ginseng in Hong Kong was about $17.57 a pound, and some unusual roots were valued at more than $130 per pound.[9]

Since 1964, the demand for American ginseng once again has increased constantly. As shown in Figure 6.2, more than $2 million worth of ginseng was exported from the United States after 1964, with increases reaching $4 million in 1966, $5 million in 1969, and $8 million in 1972. As a result of Richard Nixon's visit to China in 1972, American ginseng sales suddenly boomed in Hong Kong. China takes all of the ginseng that American growers and ginseng hunters can supply. In 1973, of the total export of American ginseng valued at about $9 million, the Chinese buyers in Hong Kong took about 94% of the total supply. Similarly in 1974, the total American

Domestic wholesale (average) price, exports, and total
value of american ginseng from 1821 to 1920, inclusive

Year	Total exported (lbs)	Total value	Average price
1821	352,992	$ 171,786	$ 0.48
1822	753,717	313,943	0.41
1823	385,877	150,976	0.39
1824	600,046	229,080	0.38
1825	475,974	144,599	0.30
1841	640,967	437,245	0.68
1842	144,426	63,702	0.44
1843	556,533	193,870	0.34
1845	468,530	177,146	0.37
1858	366,053	193,736	0.52
1862	630,712	408,590	0.84
1868	370,066	380,454	1.02
1878	421,395	497,247	1.13
1888	308,365	657,358	2.13
1889	271,228	634,091	2.33
1890	223,113	605,233	2.71
1892	228,916	803,529	3.51
1895	233,236	826,713	3.54
1897	179,573	846,686	4.71
1898	174,063	836,466	3.66
1900	160,101	833,710	5.20
1904	131,882	851,820	6.45

FIGURE 6.1   Domestic wholesale.

ginseng exported was about $11.1 million, and about $12 million in 1975.[10] The current wholesale prices are about $50–$60 per pound for wild ginseng, and about $30–$40 per pound for cultivated American ginseng at New York City's fur traders. The price fluctuates tremendously due to the sources, the age, and the quality of the root. Even as it is today, wild ginseng roots are much more highly priced than cultivated roots, because they grow naturally. All of the American ginseng exporting records were kept at the Bureau of Census of the U.S. Department of Commerce.

The United States will experience stiff competition for the ginseng market in the Orient from now on. Both Japan and South Korea, and perhaps later, China, are big ginseng growers and suppliers in Hong Kong at lower prices for better ginseng roots. In 1973 the U.S. price hit nearly $50 per pound, compared to Japan's $25 and South Korea's $35 per pound.[11]

STORY OF AMERICAN GINSENG HUNTING

Ginseng hunters and trappers frequently carried digging tools, food, etc., and went into the woods to hunt for the "green gold," and sometimes even women and boys hunted the roots. The plant was well known to all mountain children, and few were the mountain cabins that had no ginseng in them waiting for market.

Year	Total exported (Ibs)	Total value	Average price
1905	146,576	1,059,849	7.30
1906	160,949	1,175,844	7.30
1908	154,180	1,111,994	7.21
1910	192,406	1,439,434	7.48
1912	155,308	1,119,301	7.20
1913	221,901	1,665,731	7.50
1914	224,605	1,832,685	8.15
1915	103,184	919,931	8.91
1916	256,082	1,597,503	6.23
1917	198,480	1,385,203	6.98
1918	259,892	1,717,548	6.60
1919	282,043	2,057,260	7.29
1920	160,050	1,875,384	11.71
1921	181,758	1,507,077	8.29
1922	202,722	2,334,918	11.51
1923	148,385	2,245,258	15.13
1924	167,318	2,399,926	14.35
1925	138,131	1,668,221	12.07
1926	180,262	2,640,488	14.65
1927	169,000	2,556,000	15.12
1928	184,000	2,288,000	12.43
1929	234,000	2,766,000	11.82
1930	203,000	1,877,000	9.24
1931	265,000	1,922,000	7.25
1932	171,000	835,000	4.88
1933	233,400	844,000	8.62
1934	232,000	1,203,000	5.23
1935	167,000	618,000	3.70
1936	295,000	1,236,000	4.19
1937	136,000	706,000	5.18
1938	167,000	1,028,000	6.15
1964	139,206	2,731,602	19.62
1965	116,791	2,887,310	24.72
1966	173,405	4,358,542	25.13
1967	146,135	4,507,152	30.84
1968	133,701	4,359,524	32.61
1969	145,392	5,533,406	33.06
1970	162,689	5,016,951	30.83
1971	168,835	5,827,289	34.51
1972	227,549	8,922,426	39.21
1973	183,136	8,846,112	48.30
1974	216,832	11,116,787	51.27
1975*	220,000	12,000,000	54.00

* Estimated

FIGURE 6.2　American ginseng exported.

Albert Burnworth, the veteran ginseng digger, who lived near Ohiopyle, Fayette County, Pennsylvania, recalls the days of his memorable "sang'n" (ginseng hunting). The following passages are quoted from his own words published in the *Pennsylvania Angler* (https://www.fishandboat.com/Transact/AnglerBoater/LegacyIssues/1960s/Documents/11november1969.pdf):

Most people who know the stuff and who have dug it call it "sang"—and that includes me.

In front of his shaded porch, he said:

Almost everywhere hereabouts "sang" was beginning to show the effects of over-digging when I was a boy. But there were some nice patches to be found from time to time. I had no idea what the Chinese people did with "sang." I just knew that it could buy the things that people needed. In many families both girls and boys learned early to distinguished "sang" from other woods plants. Taken to a local country storekeeper the roots could be exchanged for school clothes and household needs. Shipped to the big buyers it brought money that made dreams come true. Many a boy got his first fishing tackle and his first gun and pocket watch with "sang" money.

Asked if he could think of just one old-timer who had done a great deal of "sang" digging, Mr. Burnworth replied: "Not one. They are all gone, almost like the 'sang' is gone. But there were some good 'sang' diggers in this locality. This rich Youghiogheny watershed produced some big old 'sang' roots; but personally I can't tell of digging any extremely large roots, the biggest being just a fraction under half a pound, and it was shaped like a parsnip. I reckon maybe the tales of some 'sang' roots are like a few of the stories about big fish—stretched a bit. And that reminds me, that it wasn't uncommon for 'sang' diggers of my day to get in some fishing while on a root-hunting jaunt. In fact I had line and hooks with me a good part of the time when I was 'sanging.' I have dug hundreds of pounds of 'sang' and always found the north slopes best suited to its growth. 'Sang' likes cold weather and high elevations and it almost never freezes out. I usually aimed to do most of my 'sanging' in the fall when the 'sang' berries were ripe; and all that I found I'd plant in hard-to-get-to places along the slopes of steep hollows and the like."

He continued:

I think that "sang" was struck the biggest blow during the Depression Years. A lot of country people scoured the hills for it and in a way you couldn't blame them. A pound of dry "sang" would buy a nice lot of groceries for a hungry family, or pay a doctor or buy lots of clothes; and two pounds would buy a good cow. But it's sad to think of wild "sang" being just about a thing of the past. Scarcity always does wild things to prices, and I guess that's why "sang" is now fetching about forty-two dollars a pound. I know of a couple of nice stalks of wild "sang" in a hard-to-find nook. Some of my memories make them almost sacred, so for now, I guess their whereabouts will have to be my secret. ...

REFERENCES

1. Savag, W. N., *Penn. Angler*, November 1969, pp. 18–22.
2. Time Special Bicentennial Issue, September 26, 1976, p. 44.

3. Camplin, P., "Ginseng Hunting: A Lost Art," in *Our Heritage*, Kentucky, October 1965.
4. Schorger, A. W., "Ginseng: A Pioneer Resource," *Wis. Acad. Sci. Arts and Lett.*, 57: 65, 1969.
5. Nash, G. V., "American Ginseng: Its Commercial History, Protection and Cultivation," in U.S. Department of Agriculture, Div. Botany, *Bulletin No. 16*, Revised edition, U.S. Government Printing Office, Washington, D.C., 1898.
6. Lass, W. E., "Ginseng Rush in Minnesota," *Minn. Hist.*, 41: 249, 1969.
7. Harding, A. R., *Ginseng and Other Medicinal Plants*, Emporium Publications, Boston, 1972, pp. 121–132.
8. Stockberger, W. W., "Ginseng Culture," U.S. Department of Agriculture, in *Bulletin No. 1184*, U.S. Government Printing Office, 1921, revised, 1949.
9. Williams, L. O., "Ginseng," *Econ. Bot.*, 11: 344, 1957.
10. U.S. Department of Commerce, extract from U.S. Exports, *Schedule B., Commodity by Country, EQ 629, 2928010*, "Ginseng," January to December, U.S. Government Printing Office, Washington, D.C.
11. *Agriculture Situation*, S.R.S. U.S. Department of Agriculture, Washington, D.C., May 1975, p. 12.

7 The Flourishing Ginseng Business in Korea, Japan, and Canada

THE FLOURISHING KOREAN GINSENG BUSINESS

Korea earned its fame and prestige as the ginseng country. The favorable soil and climate, particularly in the mountain areas, have provided ideal conditions for producing the highest-quality Korean ginseng roots. In the past 2,000 years, ginseng has been one of the biggest sources of income to the Korean government and its people. From 1896 to 1901, Korea exported about 7,000 *catties* of white ginseng, and about 30,000–90,000 *catties* of red ginseng to China each year. The price for white ginseng was 76 yen while for red ginseng, 2,183 yen per *catty*.[1] Before World War II, Korea exported about 26,000 *catties* of red ginseng to China each year, which earned more than $1.2 million.[2] After World War II, the Korean government continued to keep the monopoly system to manage the ginseng business. The Office of Monopoly was established in Seoul to cultivate, manufacture, and export Korean red ginseng products (Figure 7.1).

Many privately owned ginseng farms were then born to cultivate ginseng plants, making white ginseng products. As a result of systematic and industrialized cultivation in recent years, ginseng harvesting and exportation had significant growth.

In 1974 South Korea became the number one ginseng grower and exporter in the world.[3]

Korean ginseng roots used for exportation have a wide variety of specifications—that is, size, color, and age—and can be quite puzzling to buyers. Two kinds of ginseng roots are most common for export: the red ginseng and white ginseng roots. There are at least six grades or classes of Korean red ginseng roots.[3] According to the Korean government's export standard, the red ginseng classes are as follows:

Heaven Brand	First Class
Earth Brand	Second Class
Good Brand	Third Class
Large Tails	Fourth Class
Small Tails	Fifth Class
Cut Brand	Sixth Class

The current price for first-class Korean red ginseng is about $200 per 600 g retail at Chinatown, New York's drugstores.

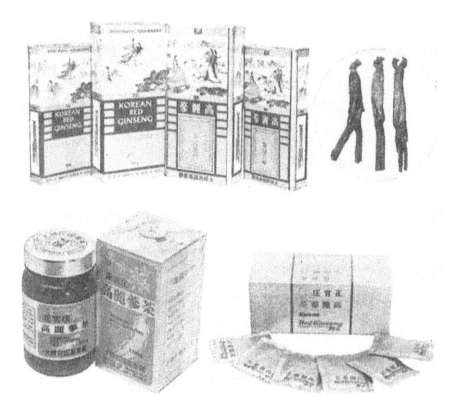

FIGURE 7.1 Korean red ginseng roots and products. (Adapted from Hou, J. P. *The Myth and Truth About Ginseng*, Cranbury, NJ: AS Barnes and Co., Inc., 1978.)

The Korean white ginseng roots are usually doubly cheaper than the red ones. Similarly, the classification is based on the age, size, and weight of the root.[3] Evidently, there are six grades of white ginseng roots, which are as follows:

Each 600 g box contains:

Class A	No more than 10 roots
Class B	Not less than 11, but no more than 20 roots
Class C	Not less than 21, but not more than 30 roots
Class D	Not less than 31, but not more than 40 roots
Class E	Not less than 41, but not more than 50 roots
Class F	More than 50 roots

The Class A white root is currently sold at about $50–$60, and the Class E grade white root is sold at about $36 dollars per 600 g, F.O.B. Korean port.

Korean red ginseng is famed throughout the world for being the highest quality, and the only competitors for international markets come from China and North Korea. The United States, Canada, and Japan are keen competitors in the sale of white ginseng root. However, since the Office of Monopoly guaranteed the quality of ginseng roots and their processed products, the exportation of red ginseng alone reached to $9.6 million in 1973, from $5.3 million in 1972, with a growth of about 82% in 1 year. Exportation of white ginseng and its products is similarly subject to government inspection (Figure 7.2).

In 1970, Japan imported more than $2 million worth of Korean white ginseng. The biggest customers of white ginseng next to Japan are West Germany and Switzerland. They usually import the top-grade white ginseng, while Japan, on the other hand, prefers the lower grade or the cheapest roots.

Currently, Korean ginseng products are shipped abroad by about 30 major ginseng industrial firms. Red ginseng is primarily exported to Hong Kong, followed by the United States, Singapore, the Federal Republic of Germany, Japan, Thailand, France,

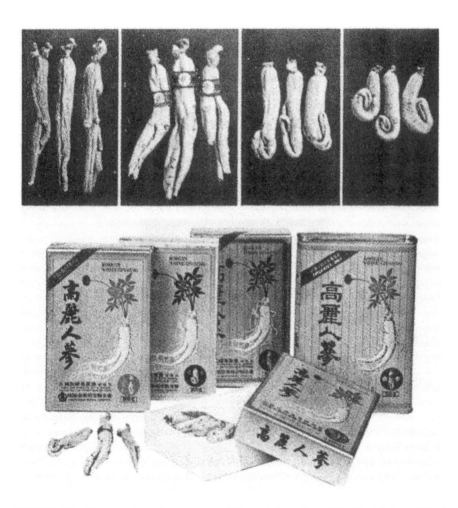

FIGURE 7.2   Korean white ginseng roots. (Adapted from Hou, J. P. *The Myth and Truth About Ginseng*, Cranbury, NJ: AS Barnes and Co., Inc., 1978.)

Italy, Holland, Canada, South America, and African countries. Korean white ginseng products are mostly shipped to Japan, Hong Kong, Switzerland, West Germany, Malaya, and the Middle East countries, according to the Office of Monopoly.

In addition to red ginseng and white ginseng roots, exported items also include ginseng in syrup or "honey ginseng," instant ginseng tea, ginseng cake and candy, ginseng extract, ginseng powder, ginseng drink, ginseng tablets, ginseng capsules, ginseng pills, ginseng fluid extract, ginseng tincture, ginseng elixir, ginseng wine, and ginseng electuary. Most of these products have been found only relatively recently in health-food stores and Oriental drugstores in the United States.

The Korean ginseng business has had records of fantastic growth. According to the Office of Monopoly, in 1970 the total amount of ginseng products exported was $8.3 million in comparison with $3.7 million in 1968, and only $492,000 in 1963. Growth was even larger as shown by earnings: about $14.2 million in 1972, $23 million in 1973, and $31 million in 1974, with a $45 million target in 1975 of total exportation of Korean ginseng products. Each year had about 30%–50% net growth, which indicates that ginseng is indeed selling very well. If the ginseng-loving population could be doubled in the next 2–3 years, one would expect Korean ginseng exportation to double, and a $100 million target would not be difficult to reach.

In recent years, red ginseng health supplements have been consumed in the forms of ginseng root, powder, tablet, capsule, concentrated extracts, soft capsule, pill, granule, beverage, candy, etc. The production amount by classification of each ginseng type in South Korea as of 2009 is estimated as follows.

In South Korea, ginseng is widely consumed by children as well as the elderly. Ginseng-consuming classes are gradually expanding to children and youth; ginseng products have become the most popular health foods for all Koreans. In South Korea, ginseng consumption stays steady throughout the year, and sales reach the highest marks around New Year's Day, Parents' Day, Chuseok holiday seasons, etc.

Red ginseng, the most attractive product of the South Korean ginseng markets, takes approximately 59% of the entire South Korean market, and the consumption continuously rises every year, growing steadily each year in every ginseng type as well as health foods.

GINSENG BUSINESS IN JAPAN

Market value of the crude cultivated *Panax ginseng* in Japan varies considerably with the sources of production and age. The Fukushima ginseng root is the cheapest, the Nagano ginseng root is medium priced, and the Shimane ginseng root is the most expensive. For example, in 1973, the 6-year-old Fukushima root was sold at 3,500 yen, Nagano root was sold at 4,500 yen, and Shimane root was sold at 6,000 yen per kilogram. From 1970 to 1973, prices went up at a rate of about 10%–20% each year as shown in Table 7.1. Accordingly, the Japanese ginseng root was sold in Japan at about $6.17 in 1970, $6.21 in 1971, $6.80 in 1972, and $7.05 per pound in 1973. The price of a 5-year-old root is cheaper than a 6-year-old root by about 10%–15%, while a 4-year-old root is cheaper than a 5-year-old root by another 10%.

TABLE 7.1

Annual Exportation of Red Ginseng, White Ginseng and Other Ginseng Products from Japan

Year		Red Ginseng	White Ginseng	Others	Total[a]
1970	Kg	80,984	4,633	2,051	
	1000¥	1,251,260	54,038	14,152	1,319,450
					($4,398,166)
1971	Kg	110,416	4,219	47,815	
	1000¥	1,875,346	49,551	12,462	1,937,359
					($6,457,863)
1972	Kg	88,185	4,094	4,537	
	1000¥	1,352,126	41,235	25,987	1,419,342
					($4,731,140)
1973	Kg	83,764	3,529	11,390	
	1000¥	1,360,572	39,436	133,596	1,534,104
					($5,113,680)
1974	Kg	109,376	4,142	13,980	
	1000¥	2,695,058	49,936	163,414	2,908,408
					($9,694,693)

[a] According to the current rate of U.S.$1 = 300¥.

The annual production of cultivated ginseng in Japan is about 319,000 kg (or 703,200 lbs) including 4-, 5-, and 6-year-old roots. Based on the market value of the different roots, the total value of Japanese ginseng is about $4.3 million.

The best ginseng roots are usually made into red ginseng for exportation. The inferior roots are made into white ginseng and other processed ginseng products consumed domestically or exported. The processed ginseng products are ginseng extract, ginseng tincture, tea, etc., sold at health-food stores.

Japanese ginseng exportation is similarly a flourishing business. In 1970, Japan exported ginseng products valued at about $4.4 million, $6.4 million in 1971, $4.7 million in 1972, $5.1 million in 1973, $9.7 million in 1974, and $10 million in 1975. Nearly 95% of the products are exported to Hong Kong, and the rest of the products are exported to Singapore, Thailand, West Germany, France, and the United States.[4]

CURRENT REGULATION OF GINSENG PRODUCTS

In the last few years, the import of foreign ginseng and ginseng products has become a boom. Almost all health-food stores, Oriental-goods stores, and even some American drugstores are now stocked with and selling imported ginseng products from Korea, the Soviet Union, and China. Ginseng tea, extracts, tablets, and capsules have been best sellers. These products are sold as food, not as drugs.

Because modern scientific information on the medicinal value of ginseng has not reached the hands of government agencies, ginseng is still banned from being marketed domestically as a drug. The U.S. Food and Drug Administration (FDA) has been effectively prohibiting the importation and marketing in the United States of all forms of pharmaceutical ginseng products bearing medical claims. The FDA similarly issued a guidance several years ago that: "We are not aware of any evidence tending to establish that ginseng should be generally recognized safe for use in alcoholic beverages or as flavoring agent in carbonated beverages or soft drinks." However, the FDA permits the import and marketing of ginseng roots and other ginseng products provided that no nutritional or therapeutic claims are made on the labels of the products.[5]

GINSENG BUSINESS IN CANADA

Ginseng has a special place in the history of Ontario and Quebec. Roots were used in traditional Native medicines. In 1715, a Jesuit priest recognized the plant from descriptions out of China and initiated export to Hong Kong. At one time, the ginseng trade rivaled the fur trade. All of the roots were harvested from the forests, and now truly wild ginseng is rare in Ontario and Quebec. In June 2008, the Endangered Species Act, 2007, came into effect in Ontario, making it illegal to plant, harvest, possess, buy, sell, lease, or trade ginseng collected from the wild in Ontario without authorization through a permit or agreement under the act.

Under the Convention on International Trade in Endangered Species (CITES), a permit is required for export of ginseng. This permit is necessary for field-cultivated roots but not for live plants, seeds, or processed roots. The exportation of wild ginseng root from Canada is prohibited. Ginseng was first cultivated in Ontario in the field under artificial shade in the late 1800s, near Waterford, Ontario. It was not until after World War II that the ginseng industry began to expand, and until the 1980s, there was a limited number of growers in Ontario. Since then, acreage has increased, and in 2010, there were more than 140 producers of this root and over 2,200 ha (5,300 acres) under cultivation.

REFERENCES

1. *Aston, Pharm. J.*, 15: 732, 1885.
2. Jones, B., "Ginseng: Seoul's Oldest Export," *New York Times*, March 14, 1971.
3. *Statistical Year Book of Foreign Trade*, Department of Customs Administration, Republic of Korea, 1974.
4. *Fujisan Marketing Report*, July, 1975.
5. *Import Alert*, U.S. Food and Drug Administration Headquarters, Field Compliance Branch January 28, 1975.

8 Current American Ginseng Dollar Value

Both American Ginseng and Asian ginseng are valued as *folk remedies* to treat everything from cancer to erectile dysfunction, but while some studies have found that ginseng may boost the immune system and lower blood sugar, there's no conclusive evidence that it can treat other medical conditions. Still, ginseng roots are highly valued, especially wild American ginseng, which Asian buyers believe to be more potent than cultivated plants.

"Wild American ginseng is considered to be the best in the world and is considerably more valuable than commercially farmed ginseng or Asian varieties," said Sara Jackson of Bat Cave Botanicals. Jackson has been growing and ethically harvesting a population of wild ginseng in western North Carolina for more than 10 years.

U.S. Fish and Wildlife reports show that exports of wild ginseng increased by about 40 percent between 2012 and 2013, with the majority of the roots going to China, where ginseng has been picked to near extinction.

Ginseng buyers in Asia pay a premium for certain types of roots. Those known as "man roots"—ones with a human shape and what appear to be body parts—can go for thousands of dollars.

One of Jackson's manroots was listed for sale on Etsy for $7,000 (Figure 8.1).

"The price of ginseng varies from year to year, but the one constant is the demand for wild ginseng roots with potency and character," she said. "This particular ginseng root is a remarkable example of a 'man root,' [which] is quite rare and sought after in the ginseng world." ...

Jackson points out that because this particular root has a feminine character and resembles a woman cradling a child, it's particularly precious, especially since ginseng is often used as a fertility aid.

However, Jackson's ginseng may also be considered valuable because of where it comes from.

Wild Mountain American Ginseng is very expensive, and 1 lb could be worth $1,600–$2,500.

Some of the most sought-after ginseng is harvested from the hills of the eastern U.S., primarily from North Carolina, Georgia, Tennessee, Kentucky and West Virginia, where ginseng hunters can find older, more valuable roots. Ginseng from these areas can sell for a few hundred dollars in summer, but by fall when the growing season comes to an end, those prices tend to rise above $1,000.

According to the U.S. Fish and Wildlife Service, the annual wholesale value of the American ginseng trade is $26.9 million.[1]

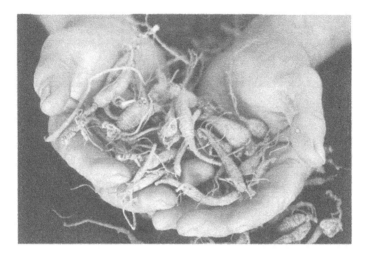

FIGURE 8.1 Wild American ginseng is widely believed to be more potent than other varieties. (Photo: Stephanie Frey/Shutterstock.)

Sales of cultivated American ginseng from Wisconsin ginseng farms, for example, sales of 5-year-old roots, depend on the sizes of the roots (Figure 8.2):

- Small-sized pieces, 8 oz about 70–80 pieces
- Medium-sized pieces, 8 oz about 45–50 pieces
- 1/4 lb, $61.00
- 1/2 lb, $121.00

GINSENG DEALERS

In the last few years, a new booming ginseng business has become prevalent. Herb shops, almost all health-food stores, Oriental-goods stores, and even drugstores now sell foreign as well as American ginseng products. You have no problem getting all kinds of modern ginseng products in almost all large cities in the United States and many countries in Europe. While it is impractical to provide a complete list of ginseng dealers in the United States, the firms listed in Table 8.1 are a few dealers that sell

FIGURE 8.2 Cultivated American ginseng from Wisconsin ginseng farms.

TABLE 8.1
Ginseng Roots and Ginseng Products Sellers

Eastern States

A A Ginseng Whol Inc Grocers
18417 Colima Road
Rowland Heights, CA 91748
(626) 810-1652

Abraham American Ginseng Company
44725 Fir Road
Gold Bar, WA 98251
(206) 794-2873

Betsy Kurtzweil
C2337 Fairview Road
Edgar, WI 54426
(715) 687-2426

Bezzie Bee
780 Redfield Road
Mosinee, WI 54455
(715) 693-3755

Bill Downey
C4217 Wiesman Road
Stratford, WI 54484
(715) 687-4437

Chuen Hing Co.
215 East 26th Street
New York, NY
(212) MU 6-5013

Clearview Ginseng
1101 South 3rd Avenue
Edgar, WI 54426
(715) 352-2997

Dale Braunel
2439 4th Street
Marathon City, WI 54448
(715) 443-2485

Dave Lemke
W6869 Cedar Street
Chelsea, WI 54451
(715) 748-5997

David Williamson
D2740 Day Avenue
Marshfield, WI 54449
(715) 384-2091

Harding's Wild Mountain Herbs
118 Walnut Street
Friendsville, MD 21531
(301) 746-5380

Heise's Wausau Farm
2805 Valley View Road
Wausau, WI 54403
(715) 675-3862

Hong Long Ginseng LLC
28 Billings Road
Quincy, MA 02171
(617) 773-0828

Hsu S Ginseng Enterprise Inc.
53 Ludlow St Front 1
New York, NY 10002
(212) 274-8653

Hus's Ginseng Enterprise
T555 Arrow Drive
Wausu, WI 54401
(715) 675-2325

Jacks Ginseng
307 North 7th Avenue
Wausau, WI 54401
(715) 845-2889

Jeff Kamenick
13554 County Road O
Merrill, WI 54452
(715) 536-9071

Jerome Seubert
R439 Lovers Lane
Athens, WI 54411
(715) 257-7296

J & K Ginseng Inc.
36911 153rd Avenue
Stanley, WI 54768
(715) 644-2733

John Rasti Ginseng Farms
3617 Golf View Drive
Wausau, WI 54403
(715) 848-7216

(Continued)

TABLE 8.1 *(Continued)*
Ginseng Roots and Ginseng Products Sellers

Eastern States

Duane Boehm
2714 West State Highway 153
Mosinee, WI 54455
(715) 693-2135

Karl Behnke
1778 County Road South
Edgar, WI 54426
(715) 443-3211

Forlin Inc.
E11101 East Bloomingdale Road
Westby, WI 54667
(608) 625-2243

K & K Roots
41 North Madison Avenue
North Vernon, IN 47265
(812) 346-7364

Gae Poong Korean Ginseng
40-15 150th Street
Flushing, NY
(212) 539-6366

Korea Red Ginseng
1580 Makaloa St. # 101
Honolulu, HI 96814
(808) 952-9966

Gary Betts
10305 SW Bryon Court
Wilsonville, OR 97070

Lucid Dreams LLC
802 East Main Street
Murfreesboro, TN 37130
(615) 200-8076

Gerald Witberler Inc.
16306 Highland Drive
Marathon City, WI 54448
(715) 443-2824

Merle Weege
4762 Sunset Drive
Mosinee, WI 54455
(715) 693-6474

Ginseng & Herb CO-OP
3899 County Road B
Marathon City, WI 54448
(715) 443-3355

Metcalf's Ginseng
412 Scott Farm Road
Afton, TN 37616

Ginseng Select Products Inc
611 Druid Road East 711
Clearwater, FL 33756

Tom Hack
2448 County Road
Mosinee, WI 54455
(715) 593-3417

Grand Wisconsin Ginseng, LLC
2402 Gowen Street
Wausau, WI 54403
(715) 432-5196

Tong Fl Enterprises, Inc.
99-17 Queens Blvd.
Flushing, NY
(212) 459-1666

Guofen Tran
6590 West 120th Avenue
Broomfield, CO 80020
(720) 524-6148

Tung Yen Tong Co.
19 Pell Street
New York, NY 10013
(212) 233-9586

modern ginseng products, with the understanding that no discrimination is intended and no guarantee of reliability is implied. Most states have lists of licensed ginseng dealers, and this information can be found on each state's website. The market prices of these ginseng products, as you may have already figured out, vary significantly as a result of source, brand, packaging, strength, etc. It is a good idea to write to or visit several different firms or stores to find out which store you think is most reasonable. Also, to shop wisely, you may check the Internet or call for lists of products and prices. Since ginseng products are sold over the counter or on the web, no prescription is needed.

REFERENCE

1. Moss, L. "Ginseng Demand Boosts Prices and Poaching." *Mother Nature Network*, MNN.com, October 2, 2015 (https://www.mnn.com/earth-matters/wilderness-resources/stories/ginseng-demand-bosts-prices-crime).

Section III

Growing Your Own Ginseng Plant

9 Growing Ginseng Plant in China and the United States

GROWING YOUR OWN GINSENG PLANT

It has been fully demonstrated that ginseng plants can be raised successfully in a field where the necessary conditions are furnished. That is to say, ginseng plants must be provided with a forest-like environment.[1-4] Those who own forest lands can grow ginseng plants easily. Also, it was formerly thought that ginseng could grow only under conditions exclusive to the Far East. However, experimental work has shown that ginseng can be raised successfully in many parts of the northern temperate zone, in some regions of European Russia and of southeastern Europe, in eastern Siberia, and in North America.[5] Ginseng can be grown from seeds, seedlings, or roots. Plants free from blight or mildew and growing spontaneously in the woodland can be transplanted to prepared gardens. Ginseng is a very slow-growing and exacting crop. It will be disappointing if not properly managed.

The cultivation of ginseng, since the beginning, has been a promising and profitable industry. As a crop, however, ginseng is no gold mine, but it certainly gives a reasonable return to growers who are willing to care for the crop for at least the 5 or 6 years that are required to reach maturity. The total acreage of ginseng farms in the whole world today is not sufficient to meet the demand of the market. Should there be a temporary decline in price, or should a glut occur, as has sometimes been the case, a grower need lose nothing, since he or she may leave the roots in the ground for 1 year or more, knowing that they are improving in size and quality. Other advantages of the ginseng industry are that it may be started and continued without excessive outlay, and may be confined to land that otherwise could not be used for cultivated crops. When properly cultivated, a small area may be made to yield a very large proportionate return.

Ginseng has become an increasingly important medicinal plant in the Western world since World War II. To meet the demand, improvements in methods of ginseng cultivation and control of typical ginseng diseases have been attained through extensive agricultural research in the Orient, the United States, and the Soviet Union. At present, increasing cultivation and output are the main concerns of ginseng-producing countries.[6]

It is most probable that the demand for ginseng will increase in the foreseeable future. The price will, naturally, continue to increase. This has been the case for the last 200 years. The large-scale industrialized cultivation of *Panax ginseng* in China, Korea, Japan, and perhaps Siberia would certainly affect the market for American ginseng. Nevertheless, the world market for ginseng has been so great that it far exceeds the supply. With more and more people turning to ginseng, the demand could double in the next decade. If you are a potential ginseng grower, you need not worry about the market.

CULTIVATION OF GINSENG IN THE ORIENT

Growing ginseng is a rather difficult job because it does not grow in conditions different from its natural environment. Ginseng grows most happily in the northern temperate zone between 36° and 38° north, where the degree of humidity is exactly suited to its prima donna temperament. At all stages in its development, the young ginseng plant must be protected from snow, rain, frost, hurricane, and, most important of all, direct sunlight. The plant is covered with thatched shades like terraced rows of little bus shelters. The ginseng plant, by nature, is a dark lover, growing without the need for too much sunshine or radiant energy. Accordingly, receiving the right amount of light is important during its growth.

The root is not harvested until it has reached the sixth year, and by that time it has sucked nearly all of the natural nutrients from the earth so that the field cannot be reused for another ginseng crop for about 10–15 years. Top-quality soil produces superior ginseng. For the seedbed, a soil of sand or gravelly sand formed through weathering of granite, gneiss, and calcite is recommended. For the field, soils composed of a texture of sandy clay loam, sandy loam, and loamy sand are suitable. All types of chemical fertilizers are found unsatisfactory or even harmful for growing ginseng plants. Special composites, called *yakto*, are used to increase the high quality of the root. The *yakto* is a complex product made from fermenting the raw foliage of broad-leaved trees and a small amount of green grass, cottonseed cake, soybean cake, and defatted rice bran. Fermented human waste mixed with animal waste, tree ashes, and chimney soot has also been used. In both the seedbed and the field, *yakto* or natural fertilizer is usually employed as the prime source of nutrients. The top-grade bed soil is made with sufficient *yakto* and chimney soot. The nutrients absorbed by the plant thus stimulate its growth. The quality of the ginseng root, no doubt, depends on the soil and nutrients used during cultivation.

Seeds are taken from 4-year-old plants and raised to seedling stage in carefully drained seedbeds. The season for sowing seeds of ginseng is preferably in late October or early November. The seeds grow in a nursery for one and a half years, and in March or April they are transplanted into the field or permanent beds. Usually, the seedlings are planted out in beds raised a foot above the level of the surrounding soil, bordered with upright slates, and covered from heavy sun and rain by shades of reeds 3- or 4-feet high, well closed in, except toward the north side, where they are left more or less open, according to the weather.

Sheds are placed in rows with just enough room to walk between them. Ginseng is transplanted frequently during this period. Ginseng rarely blooms in less than 2 years, older plants start flowering around the middle of May and bear red fruit in middle or late July.[15] The stem is about 6 inches high with four horizontal leaves standing out from the stem at right angles, and in the fifth year, a strong, healthy plant has reached maturity, though it is usual not to take it up until it has reached the sixth year. The roots, after harvesting in autumn, are very carefully washed and scraped, and are then treated in one of two ways traditionally used by the Chinese (see Chapter 2), yielding white or red ginseng roots, respectively.

The ginseng seed has a hard coat on its surface, and since the embryo is incomplete, some manipulation is necessary to accelerate its germination. Methods to accelerate germination, control of typical ginseng plant diseases, and the interrelation between the raising process and the environment have been under intensive research in the Orient. The results of these types of studies are covered in the following sections.

GROWING GINSENG IN THE UNITED STATES

Ginseng grows naturally on slopes of ravines and in well-drained sites where soil is formed from the acid leaf mold of hardwood forests. The soil should be naturally dry and fairly light, and in a condition to grow good vegetables without the addition of strong manure. By proper treatment almost any fairly good soil can be conditioned for ginseng growing. The addition of woodland soil tends to produce hard, flinty roots of inferior quality.

In the book called *Ginseng and Other Medicinal Plants*, written by A.R. Harding in 1908, detailed information on the habits and cultivation of American ginseng is given.[4] The Farmers' Bulletin, Nos. 1184 and 2201, published by the U.S. Department of Agriculture, also give detailed information on growing American ginseng. The following information on growing and harvesting American ginseng given in this chapter is based on the *Farmers' Bulletin,* Nos. 1184 and 2201, respectively.

First, American ginseng must be provided with favorable conditions for growth. Selection of the proper location, preparation of the soil, and good drainage are important in planting ginseng. The best site for beds is a hardwood forest, with tall trees to provide favorable dense shade, and with little undergrowth. Similar drainage and shade conditions should be maintained when growing ginseng in lath sheds. Make beds 4 feet wide with walkways between them. For root planting work the beds up to 12 inches deep. For seeds and seedlings, work the beds only 8 inches or so deep to prevent settling. Mound the center of permanent planting beds to provide space for more plants and, if located on flat ground, to facilitate good runoff of water. Slope the walkways so that they will drain water from the beds during heavy rains. The most favorable conditions for the soil are a rich, sandy, loamy soil. Clay land can be used, but one has to mix it with leaf mold, rotten wood and leaves, and some light soil, and it must be thoroughly pulverized.

Ginseng needs three-fourths shade during the summer and free circulation of air. The proper amount of shade can be provided in lath sheds or by trees in a forest planting. Laths should run north to south to provide alternating sun and shade to the plant. Do not use burlap or muslin; they interfere with air circulation.

For seedbeds, break up soil to a depth of 6–8 inches, and remove all weeds, grasses, and roots. Mix one-to-one with fiber-free woodland soil. If the soil is inclined to be heavy, add enough sand so that the mixture will not harden after heavy rain.

Use of seeds instead of seedlings may prevent the introduction of disease to new plantations. Also, this is the least expensive way to start a plantation, but it requires a longer period until harvest.

Seeds ripen in the fall but generally do not germinate until the following fall. Do not allow ripe seeds to dry out. Store them in a cool, moist place. Use woodland soil, sand, loam, or sawdust as a storage medium.

Plant seeds in the spring, as soon as the soil can be tilled. Only scarified or partially germinated seeds should be used for planting. They are planted 8 inches apart each way in permanent beds, or 2 × 6 inches apart in seedbeds. Cover seeds with 1 inch of forest soil, or well-rotted or basswood sawdust; do not use pine or oak sawdust.

Some growers plant the seeds when they ripen in September and cover the beds with leaf mold or mulch. They keep the beds covered until spring, when the seeds begin to sprout.

Ginseng seedlings are more expensive than seeds, but a crop grown from seedlings can be harvested 2 or 3 years sooner than a crop propagated from seeds. Several firms sell 1-, 2-, or 3-year-old seedlings. Three-year-old seedlings produce seed during the first fall after planting, which may be used for planting future crops. Set seedlings in permanent beds, 8 inches apart each way. Closer spacing tends to increase disease in the plantation.

Roots may be set any time from October to April, after the soil has been tilled. Fall planting, however, is usually preferred. Plant roots 2 inches below the bed surface and 8 inches apart each way. When roots are not available from woodlands, beginners should purchase them from reputable growers. Roots grow more rapidly when not permitted to seed.

Ginseng may be grown directly in woodlots, or in lath sheds with partial shade—an environment similar to the plant's natural habitat. Plants thrive best in loamy soil, such as found in oak and sugar maple forests in the North. Shade is essential.

Ginseng requires relatively little cultivation. The beds should be kept free of grass and weeds, and the soil should be scratched with a light implement whenever it shows signs of caking. One active man can easily take care of about 2 acres of ginseng.

A winter mulch over the crown is essential to prevent heaving by frost. A 4- or 5-inch layer is ample in the most severe climate; less is needed in the South. Spread mulch when frost is imminent, and remove it in the spring before the first shoots appear. Light mulching to retain moisture during dry weather is also advisable. Forest leaves or light brush, held in place with poultry netting, makes the best mulch. Cornstalks stripped of husks, bean vines, cowpea hay, and buckwheat straw are also suitable if they do not contain weeds, seeds, or other material attractive to rodents.

Many growers are opposed to excessive use of fertilizers. Heavy use of barnyard and chemical fertilizers lessens the resemblance of cultivated ginseng to the wild root. Overmanuring also forces growth and lowers the resistance of ginseng to the attack of disease.

Some growers fertilize with leaves or old sawdust from hardwood trees, or with ground-up, rotten hardwood. Others prefer woodland soil or rotted leaves 4–6 inches deep, spaded to a depth of about 8 inches, with fine, raw bonemeal well worked in, and applied at the rate of 1 pound per square yard.

Fence beds keep out animals and discourage theft. Protect the beds from moles with boards or close-mesh wire netting set 12–18 inches in the ground. Rodents may be controlled with traps.

A ginseng crop matures in 5–7 years. Generally the roots are dug in mid-October of the sixth year. Good roots are about 4 inches long, 1 inch thick below the crown, and average one ounce in the fresh state. Older roots possess the most substance and when properly cured bring the highest prices.

The proper time for digging ginseng roots is in autumn, mostly in October, and they should be carefully washed, sorted, and slowly dried. If the ginseng roots are collected at other seasons of the year, they will shrink more and not have the fine, plump appearance of the fall-dug root. Dig the roots with their forks intact. Carefully free them of adhering soil so as to preserve their natural color and characteristic circular markings. Do not scrape or scrub them. The market value of the product is based, in part, on wholeness and appearance. Some growers replant young and undersized roots, or heel them in until spring planting.

The clean, fresh roots are usually dried in a well-ventilated heated room, at about 60°F–80°F, and after a few days, the temperature of the room can be raised up to 90°F. until the roots dry. Spread the roots thinly on lattice or wire-netting shelves. Turn them frequently but handle with care to avoid marring the surface or breaking the small branches. Roots more than 2 inches in diameter will need to be dried for about 6 weeks. During damp and very wet weather, care should be taken to see that the roots do not mold or sour. They should never be overheated, since this will tend to discolor the surface and spoil the texture of the roots. Too fast drying at too high a temperature damages the roots both physically and chemically. When all cured, the roots should be stored in a dry, airy, and rodent-proof place or in containers.

It is of great importance that the roots should be properly treated for marketing. They should never be split in washing or drying. The little neck or bud-stem should be unbroken, for if it is missing, the roots lose two-thirds of their value in the market. In the ginseng business, as in other trades, there are tricks. The tricky ginseng diggers have been known to try to adulterate the ginseng shipments with pokeweed and other roots. The inexperienced buyer may at times be fooled, but not the dealer who really knows his business. Ginseng roots have also been doctored with thin slivers of lead and other weights to make them heavier. Sometimes, depending on how skillfully the insert is made and camouflaged, the deceptive scheme is difficult to detect. Of course artificial weight has to be added when the ginseng is green, and sometimes the drying process cracks the roots and exposes the petty fraud. No matter how well it grades out, Chinese buyers never rate American ginseng better than *third* class.

The cultivation of American ginseng and the control of ginseng diseases have been researched by both governmental agencies and many Agriculture Experiment Stations at state colleges and universities during the ginseng boom years. In 1895, the U.S. Department of Agriculture published *Farmers' Bulletin* (no. 16) entitled "American Ginseng, Its Commercial History, Protection, and Cultivation." The request for this bulletin was great enough to require the revision and printing of a new edition in 1898. In 1902, the bulletin was again reprinted with the addition of a "Note of Warning" signed by Frederick V. Coville calling attention to a "BOOM" in the sale of ginseng seeds and roots. Because of the extravagant prices, fraudulent and adulterated species were unavoidable. In 1904 and after, bulletins on ginseng culture and diseases were also published by state agricultural experiment stations in Pennsylvania, New York, Kentucky, Missouri, and possibly elsewhere. In 1913, a new ginseng bulletin (no. 551) written by V.F. Valter, entitled "The Cultivation of American Ginseng," was published. In 1921, the U.S. Department of Agriculture issued another bulletin, *Farmers' Bulletin* no. 1184, edited by W. W. Stockberger, entitled "Ginseng Culture" and reprinted in 1941 and 1953. The most recent issue of a ginseng bulletin, the *Farmers' Bulletin* no. 2201, written by L. Williams, entitled "Growing Ginseng" was issued in 1963. The requests were so enormous that the same bulletin had to be reprinted in 1964 and 1973.

We do not know how many Americans are still interested in growing ginseng plants today, but there are many. A recent article in the *New York Times*[5] described how Denver Davis grows his ginseng plants in the hills of northern Georgia. Davis started growing ginseng as a hobby 19 years ago. He said: "When I first started, the first year I got a matchbox full of seeds, and I thought I had done some good. Now I am digging 12 bushels a year, getting two gallons of seeds." On his 2-acre ginseng farm, the ginseng plants are from 1 to 15 years old, and most of them are mature and ready to harvest. Once the plants are mature, the tops are discarded, the roots are harvested, and the roots are dried in a heated building like a tobacco barn. They lose two-thirds of their weight in moisture before the pale-yellow roots are ready for sale. Another ginseng grower, Mr. Gooch, also of Georgia, owns a 4-acre ginseng farm. He harvests 300–600 pounds of ginseng root per acre, worth $20–$30 a pound to dealers, while the dealers, usually fur and hide traders, sell at as high as $60 a pound to the Asian buyers. Talking about growing your own ginseng plant, Davis said: "You have to week it five to six times a year, the moles and rats are after it all the time. A lazy man ain't going to grow it."

SPEEDING UP GERMINATION AND GROWTH

Since in the newly ripened seeds the embryo is not yet formed, the sprouting of ginseng seeds usually takes about 18–22 months. The development of the embryo, under natural conditions, takes several additional months after ripening of the seeds. This fact was revealed only a few years ago, but the Chinese and Korean ginseng growers empirically knew about this and worked out special pre-sowing treatments of the seeds. However, their traditional methods proved not to be satisfactory according to the standards of modern agriculture.[6]

Seeds stratified immediately upon ripening sprouted after 8 months; those stratified after 4 months of storage under dry conditions sprouted in 19 months. In the last few years, Japanese and Soviet botanists have developed a new and effective method of hastening germination with a chemical agent called *gibberellic acid*.[6] Gibberellic acid has been widely used in promoting growth of plants, especially the growth of seedlings.[7] Gibberellic acid belongs to the family Giberellins, and so far at least 14 gibberellins have been isolated. This plant growth promoter is obtained from the fungus *Gibberella fujikoroi* (Sawada) Wollenweber.

Soviet botanists Grushivitskii and Limari found that in the ginseng seeds treated with gibberellic acid, the length of the first stage of after-ripening is reduced from 4 to 2 months; consequently, the whole period of preparation of the seeds for sowing is reduced from 8 to 6 months.[8] According to Japanese agriculturalists Ohsumi and Miyazawa, when ginseng seeds were soaked in various concentrations of gibberellic acid solution and incubated in a sand bed, gibberellin promoted the growth of the embryo and, as a result, raised the rate of germination.[9] The germinating power, or the number of seeds germinated, was increased approximately from 50%–70% to 90%–100%. The best results were obtained if the seeds were previously treated with 0.05%–0.1% of gibberellic acid solution over a period of 25 hours.[10]

Other germination-promoting agents, such as kinetin, naphthal-epeacetic acid, indole-3-acetic acid, 2,4-D in addition to gibberellic acid were also tested. The best growth-promoting agent so far detected, was gibberellic acid. Recent studies in Japan used a 100 p.p.m. aqueous solution of gibberellic acid in which seeds were immersed for 24 hours, and germination of seeds was accelerated by lowering the temperature to 2°C–15°C for about 10 days. The optimum temperature for germination was found to be 10°C.[11]

The effects of light intensity and pH of the culture medium on ginseng plant growth have been reported by Kuribayashi and associates.[12,13] Two-year-old plants were grown in Wagner sand pots and supplied with Hoogland and Arnon nutrient solution. The light intensity was adjusted to 100%, 50%, 30%, 10%, and 5% of natural normal sunlight, and the pH of the soil was adjusted to 3, 4, 5, 6, 7, and 8, respectively. After 4 months of cultivation, it was found that no plant survived at the 100% and 50% light intensities, whereas the majority of the plants survived at 5%–10% sunlight (about 3,000–6,000 lux) intensity. High alkalinity destroys ginseng plants. Many ginseng plants died at the environment soil of pH 7 and 8, whereas they grow normally in pH 4, 5, and 6 medium. Accordingly, 5%–10% sunlight intensity, and the slightly acidic environment medium with pH of 5–6 are the most optimum conditions for ginseng growing.[14]

Research also indicates that light has a significant effect on the absorption of the nutrient by the plant. The concentrations of the plant nitrogen, phosphor, and potassium are different under different light intensity conditions. The maximum rate of intake of these nutritive substances by the plant is under shade or dark conditions. The common forest soil on which ginseng normally thrives is usually brownish-gray in color, containing a relatively greater amount of aluminum, calcium, magnesium, nitrogen, organic matters, and sulfur. Plant research shows that soil with a relatively higher content of sulfur, magnesium, calcium, in addition to the other three essential nutrients, is very important for promoting the normal growth of ginseng culture.

REFERENCES

1. Hong, S. K., in *Korean Ginseng Science Symposium*, Korean Society of Pharmacognosy, Seoul, 1974, pp. 37–54.
2. Stockberger, W. W. "Ginseng Culture," in *Farmers' Bulletin No. 1184*, U.S. Department of Agriculture, 1921, Revised 1941.
3. Williams, L. "Growing Ginseng," in *Farmers' Bulletin No. 2201*, U.S. Department of Agriculture, 1964, Revised 1973, U.S. Government Printing Office, Washington, DC.
4. Harding, A. R. *Ginseng and Other Medicinal Plants*, Emporium Publications, Boston, 1972.
5. King, W., "Georgean Farm," *The New York Times*, October 11, 1975.
6. Baranov, A. *Econ. Bot.*, 20: 403, 1966.
7. (a) Bukovac and Wittwer, *Q. Bull. Michigan Agric. Exp. Stn.*, 39: 307, 1956. (b) Merritt, *J. Ag. Food Chem.*, 6: 184, 1958.

8. Grushvitskii, I. V. and R. S. Limar, *J. Bot.*, USSR, 50: 215, 1965.
9. Osumi, T. and Y. Miwazawa, *Nogyo Oyobi Engei*, 35: 723, 1960, via *Chem. Abstr.*, 56: 2731g, 1962.
10. Nikolaeva, M. G., I. V. Grushvitskii, and V. M. Bogdanova, *Bot. Zh.* (Leningrad), 57: 1082, 1972, via *Chem. Abstr.* 78: 25215x, 1973.
11. Kuribayashi, T., M. Okamura, and H. Ohashi, *Shoyakugaku Zasshi*, 25: 87–94, 1971.
12. Kuribayashi, T. and H. Ohashi, *Shoyakugaku Zasshi*, 25: 95, 1971.
13. Kuribayashi, T., M. Okamura, and H. Ohashi, *Shoyakugaku Zasshi*, 25: 102, 1971.
14. Kuribayashi, T. and H. Ohashi, *Shoyakugaku Zasshi*, 25: 110, 1971.
15. Maisch, J. M., Ed., *American Journal of Pharmaceutical Education*, 57: June, 1885.

10 American Ginseng Farm in the State of Wisconsin

Wild American ginseng is generally found in northern and northeastern areas of the United States. Cultivation is done in many states, such as Kentucky, North Carolina, Tennessee, Wisconsin, and many others. Today, Wisconsin ginseng farms account for 95% of the total cultivated ginseng production of the United States, with approximately 85% of it exported overseas. American ginseng roots grown in Wisconsin contain very high levels of ginsenosides, the active ingredient in ginseng, compared to American ginseng grown elsewhere. It is therefore highly sought after. Wisconsin American ginseng can be recognized by a trademarked seal that ensures the product is of the highest Wisconsin quality.

BRIEF HISTORY OF CULTIVATED GINSENG IN MARATHON COUNTY, WISCONSIN

The excitement behind the cultivation of American ginseng began more than 110 years ago with the Fromm brothers in the small community of Hamburg, Wisconsin. These industrious brothers introduced Marathon County's unique crop to the world in the early 1900s. Marathon County, Wisconsin, grows the most prized and demanded American ginseng in the world.[1]

Hsu Ginseng Farms utilize state-of-the-art cultivation techniques and equipment with unprecedented investment in facilities, research, and testing. The whole process of growing American ginseng begins with preparation of virgin ginseng land and moves through a process of the stratification of seed, planting, care, and harvest. The care of a ginseng plant is very labor intensive and requires a huge investment.

American ginseng can only be grown on the land once. Contrary to popular belief, the plant does not remove anything from the soil. In the wild a ginseng plant can live for decades, so Mother Nature has given the plant the capability to protect itself from being smothered out due to the fact it can drop 15–20 seeds every year. Due to the density of seed at which we plant, the soil cannot handle the growth of another ginseng plant. From planting to harvest can take as many as 5 years in a cultivated setting of artificial shade. Our crews can hook as many as 600,000 hooks a year to hold the shade covers in place, only to remove many of them prior to winter or harvest.

Hsu Ginseng Farms employs highly qualified staff and seasonal workers, more than 100 during harvest season. Their state-of-the-art farm facilities consistently store, process, and dry this valuable crop. At the point of harvest, the plant becomes a commodity that Hsu's Ginseng Enterprises sells to the international marketplace.

It is estimated that Wisconsin grew 3,000–5,000 acres of ginseng in 1990, and sales of the root earned almost $70 million for farmers in Marathon County. Most of Wisconsin's ginseng growers cultivate no more than 1 acre of the crop annually.

Most of the nation's ginseng crop is exported to Hong Kong, where it enters duty-free. Much is then redistributed to other locations in the Far East.

Ginseng can be a profitable crop, but it requires an enormous commitment of time, money, and labor for successful commercial production. Ginseng beds in Wisconsin are usually cultivated for 3 years before harvest, unless disease problems mandate earlier harvest.

The following sections explain the growing of ginseng in Wisconsin.

GROWTH HABITS

American ginseng plants are generally started from seeds. Seedlings or roots for transplanting are available commercially but are used infrequently. Seeds are planted in the fall and germinate in the spring. Although researchers have examined ways to break this juvenility requirement and hasten germination, it is still not understood.

First-year seedlings produce one compound leaf with three leaflets. This leaf, 1–2 inches in height and spread, is the only above-ground growth in the first year. Underground, the plant develops a thickened root about 1 inch long and up to 1/4 inch wide. At the top of the root, a small rhizome or "neck" develops with a regeneration bud at the apex of the rhizome. In autumn, the leaf drops, and a stem supporting new leaves emerges from the regeneration bud the following spring.

The plant develops more leaves, with more leaflets, each year until the fourth or fifth year. A mature plant is 12–14 inches tall and has three or more leaves, each consisting of five ovate leaflets. Leaflets are approximately 5 inches long and oval shaped with serrated edges. In midsummer, the plant produces inconspicuous greenish-yellow clustered flowers. The mature fruit is a pea-sized crimson berry, generally containing two wrinkled seeds.

After 3 years of growth, the roots begin to attain a marketable size (3–8 inches long by ¼–1 inch thick) and weight (1 oz). In older plants, the root is usually forked. Wild or high-quality cultivated ginseng root has prominent circular ridges. The highest quality mature root breaks with a somewhat soft and waxy fracture. Young and undersized roots dry hard and glassy and are less marketable.

ENVIRONMENT REQUIREMENTS

Climate

Ginseng grows best under conditions that simulate its natural habitat. It requires 70%–90% natural or artificial shade. Ginseng thrives in a climate with 40–50 inches of annual precipitation and an average temperature of 50°F. It requires several weeks of cold temperatures for adequate dormancy.

Soil

Ginseng generally prefers a loamy, deep (12 inches), well-drained soil with a high organic content and a pH near 5.5. Extremely sandy soil tends to produce long, slender roots of inferior quality.

SEED PREPARATION AND GERMINATION

Most ginseng crops are started from seed, rather than roots or seedlings. This is the least expensive way to start a plantation and may help prevent the introduction of soil-borne disease to new plantations. Ginseng requires 3–5 years to produce a marketable crop from seed.

As there is an 18-month seed dormancy, freshly harvested seed cannot be used for starting a crop. It must be stratified for 18–22 months before planting. Seed stratification involves soaking the seed in a formaldehyde solution and in a fungicide, then burying the seed outdoors in moist sand. Most seed is already stratified when it is purchased and needs only to be treated with a fungicide and sown. Seed should not be allowed to dry out before or after seeding. (For detailed instructions on seed stratification, see Reference [2]).

CULTURAL PRACTICES

SEEDBED PREPARATION

For planting seeds or seedlings, till the soil to a depth of 8–10 inches, and remove rocks. For root planting, work the beds 12 inches deep. For best results, mix soil in 1-to-1 with fiber-free woodland soil. Make beds 4 ft wide with alleys between them for walkways and for farm equipment. If the bed is on flat ground, mound the center to facilitate good runoff. Slope the walkways so they will drain water from the beds during heavy rains.

Shade can be provided by wooden lath sheds or polypropylene fabric. Artificial shade should be placed about 7 ft above the ground to ensure good air circulation. Do not use burlap or muslin, which can interfere with air circulation. (For more detailed instructions on how to provide artificial shade, see "American Ginseng Culture in the Arid Climates of British Columbia" by Oliver, Van Lierop, and Buonassisi).

SEEDING DATE

Ginseng seed is generally planted in the fall and covered with mulch until spring. It can also be spring-planted, but if seeding is not completed by May 1, the seed may begin to sprout prematurely.

Roots can be transplanted any time after the tops of the plants have begun to die back but before the ground has frozen.

METHOD AND RATE OF SEEDING

Plant seedlings 1/8 to 1/2 inch deep and 4 inches apart in the row. Space the rows 6 inches apart across the bed. The recommended seeding rate for a 4 ft wide bed with 2 ft side paths between beds is 80–100 lb/acre. To keep the seed from drying out, the beds should be covered immediately with 2 to 3 inches of straw.

Plant roots at a 30°–45° angle from the vertical, with the crown of the root ¾–1 inch deep. Cover the bed immediately with 1–2 inches of straw. A 4–5 inch layer of mulch is necessary on fall transplants to prevent heaving in frost. Some of the mulch can be removed in the spring before the first shoots appear.

Set seedings 8 inches apart in each direction. Closer spacing tends to increase disease in the plantation.

Light mulching (1–2 inches thick) to retain moisture during dry weather is advisable.

FERTILITY AND LIME REQUIREMENTS

Heavy use of manure or commercial fertilizers lessens the resemblance of cultivated ginseng to the wild root and hence may reduce marketability. Overmanuring may also force growth and lower disease resistance. Although little research in ginseng fertility has been conducted, common practice has been to fertilize as for other root crops. Recommended rates are about 15 lb P_2O_5/acre and 60 lb K_2O/acre for soils testing in the optimum range for vegetables (30–45 ppm Bray P1 and 140–200 ppm soil test K).

Nitrogen needs a range from 20 to 60 lb/acre, depending on soil organic matter level. (However, some growers have been known to use considerably more.) Growers have tended to use lower-salt fertilizers, such as ammonium sulfate, potassium sulfate, and potassium-magnesium sulfate. Although secondary and/or micronutrients are often involved in fertilization programs, little research has been conducted to confirm responsiveness.

Some growers fertilize with leaves or old hardwood sawdust or with ground-up rotted hardwood. Others prefer woodland soil or rotted leaves 4–6 inches deep, spaded to a depth of about 8 inches with fine raw bonemeal (1 lb/sq. yd.) worked in.

Fertilizers should be applied during the dormant season at least a couple of weeks before plants emerge (Figure 10.1).

VARIETY SELECTION

Although no improved varieties have been developed, American ginseng shows variations in certain characteristics, particularly in the roots. Plants from the northern part of the country, particularly Wisconsin and New York, are considered good breeding stock, because they furnish roots of good size, weight, and shape.

WEED CONTROL

Weeds can be controlled mechanically with mulching and hand weeding and chemically with Fusilade 2000. See Table 10.1 for instructions on herbicide use.

DISEASES AND THEIR CONTROL

Ginseng is susceptible to a number of fungal diseases, including Alternaria leaf and stem blight, *Phytophthora* root rot and foliar blight, seedling damping-off caused by *Pythium* and *Rhizoctonia*, rusty root, and root knot nematode. Ginseng gardens that are cultivated in the woods may suffer less from diseases than do plantings under artificial shade.

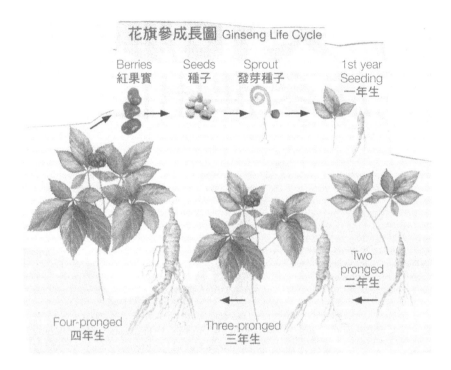

FIGURE 10.1 Hsu Ginseng Farms. (Adapted from Hou, J. P. *The Myth and Truth About Ginseng*, Cranbury, NJ: AS Barnes and Co., Inc., 1978.)

To minimize disease problems, select a growing site with good drainage. Good air circulation is also crucial and can be attained by providing cleared areas (walkways) around the beds, relatively uncrowded spacing, and control of weeds. Thin spacing also reduces the likelihood of disease spread through foliar or root contact. Wisconsin growers generally do not reuse a ginseng field for succeeding ginseng crops.

Table 10.1 shows pesticides labeled for nationwide use on ginseng. The University of Wisconsin has obtained approval from the Environmental Protection Agency (EPA) (under Sections 18 and 24 of the Federal Insecticide, Fungicide and Rodenticide Act) for the special use of several additional fungicides. Approval is granted for use in Wisconsin only, and use must be reported to the Wisconsin Department of Agriculture. Consult your local county extension agency each year to find out which pesticides may be applied to ginseng in your area.

INSECTS AND OTHER PREDATORS AND THEIR CONTROL

Ginseng is sometimes attacked by white grubs and wireworms. Voles and field mice may feed on the roots. See Table 10.1 for recommended pesticides.

HARVESTING

In Wisconsin, most growers harvest ginseng the third year after planting from seed. The roots are dug in the fall and vigorously washed to remove surface soil.

TABLE 10.1
Pesticides labeled for nationwide use on cultivated ginseng as of November 1, 1991

Pest	Materials	Treatment(s)	Restrictions/Comments
Alternaria Left and stem blight	Tankmix *Rovral* 50 W or *Rovral* 4 F (Rhone-Poulenc)	2 lbs/acre	Best if applied 8 hours before rain; *do not* apply *Rovral* within 36 days of harvest
	Champion *wettable* *powder* Hydroxide (Agtrol Chemical)	2.6 lbs/acre in min. 100 ga/acre	Also available as *flowable champ.* Tankmix 3.5 pts/acre 2 lbs *Rovral* 50 W. *EPA reg no:* 55146-1
Phytophthora Root rot and/ or foliar phytophthora	No materials labeled for use on ginseng nationwide		

Insects

Pest	Materials	Treatment(s)	Restrictions/Comments
(Soil)	*Diazinon* 14 G (Ciba-Geigy)	14–28 lbs/acre preplant and incorporate to 4–8″ depth 21 lbs/ acre broadcast over beds (spring, summer, or fall)	One preplant and one broadcast treatment/year in first and second seasons only; recommended broadcast just before rain; *do not* apply within 1 year of harvest. *EPA reg no:* 100–46 g
(Above ground)	*Pyrezone* *crop spray* (Fairfield American)	Up to 12 oz/acre	A broad-spectrum contact spray. *EPA reg no:* 4816-490
	Diazinon AG 500 (Ciba-Geigy)	0.75–1 pt/acre	No more than 1 application/year; *do not* apply during flowering on 3- and 4-year-old plants. *EPA Reg No:* 100-461
Slugs	Deadline bullets (Pace National)	20–40 lb/acre; apply at 3- to 4-week intervals as needed	Follow all label instructions for storage, application, and disposal of this product. *EPA Reg No:* 8501-34
Weed control	Fusilade 2000 (ICI Americas)	1 qt Fusilade 2000/acre plus 1% crop oil (e.g., 1 gal/100 gal), or 0.25% surfactant (e.g., 1 qt/100 gal)	Apply when grasses are 2–8″ tall, before tilling or heading; direct spray away from ginseng foliage; *do not* apply within 1 year of harvest. *EPA Reg No:* 10182-104

Source: Compiled by Jennifer Parke and Brian Hudelson, University of Wisconsin–Madison.

Note: Use only approved materials. Follow label directions. All recommendations are in terms of product per acre (not a.i. [active ingredient]).

It is important to handle the roots carefully to keep the branching forks intact and maintain the natural color and circular markings.

DRYING AND STORAGE

Ginseng roots are dried on wire-netting shelves in a heated, well-ventilated room. Since overheating destroys color and texture, begin drying the roots at a temperature between 60°F and 80°F for the first few days, then gradually increase it to about 90°F for 3–6 weeks. Turn the drying roots frequently. Store the roots in a dry, airy, rodent-proof container just above freezing.

YIELD POTENTIAL AND PERFORMANCE RESULTS

Yields of dried roots from a well-managed planting average about 1 ton/acre, although greater yields are often reported.

A typical seed yield is 150–250 lb/acre.

ECONOMICS OF PRODUCTION AND MARKETS

Ginseng growers typically invest $20,000/acre and 600 hours of labor annually and get no return on their investments until the third or fourth year. Seed and shading materials alone can cost more than $29,000/acre. It may take 10 years to break even. An average crop might net $30,000/acre, depending on the price, which tends to fluctuate widely from year to year. Prices for dried roots range from $20 to $45/lb. Seed sells for $50 to $100/lb.

In Wisconsin, growers are assessed $0.15/lb of dried root for promotion and research, and the funds are administered by the Ginseng Board of Wisconsin, located in Wausau, Wisconsin. There are several seed and root suppliers and ginseng buyers in Wisconsin. For information, contact the Ginseng Board of Wisconsin, or the Wisconsin Ginseng Growers Association (both at 500 3rd St., Suite 208-2, Wausau, Wisconsin 54401; phone 715-845-7300).

REFERENCES

1. Departments of Horticulture, Plant Pathology, and Soil Science, College of Agricultural and Life Sciences and Cooperative Extension Services, University of Wisconsin, Madison, WI.
2. Oliver, Van Lierop, and Buonassisi. American Ginseng Culture in the Arid Climates of British Columbia.

Section IV

Data and Research

11 Ginseng Information and Scientific Research

AGE-OLD UNPOPULAR PANACEA

The age-old panacea tonic is not popular in Western countries, particularly in the United States. At the present time, little modern scientific information about ginseng is available, and the majority of people, even health-care professionals, know little about ginseng.

Recently, I inquired of 33 information librarians at 24 good-sized public and drug-companies' libraries in the New York metropolitan area about ginseng. "Ginseng? I never heard of it." After searching their filing cards and reference books they said, "Sorry, we don't have any of the latest scientific information about ginseng." Twenty-six librarians gave me the same answer. Seven librarians, however, told me: "Ginseng, it is an Oriental tonic. It gives you power and vitality." Two among them not only knew about ginseng, but they told me that they are currently using it. Regardless, no single library could provide me with any scientific references. Instead they showed me some obsolete legends and tales about ginseng found in several booklets or encyclopedias about medical plants and herbs.

From 1971 to 1976, only three out of ten larger newspaper publishers printed articles on ginseng. *The New York Times* published three articles: "Ginseng—Seoul's Oldest Export" appeared on March 14, 1971[1]; "Bars Sale of Sex Stimulant" appeared on April 25, 1972[2]; and "Ginseng Root Is a Minor Cash Crop for Georgians" appeared on October 11, 1975.[3] The *Wall Street Journal* reported an article: "Lowly Ginseng Plant Is One Root to Profit for Some Americans," appearing on September 9, 1975.[4] The *Milwaukee Journal* reported two articles: "Interest in Growing Ginseng Root Renewed," and "Ginseng Still has a Lot of People Rooting For It," appearing on April 8, 1973 and December 13, 1974, respectively.[5,6] The other seven newspapers I surveyed (e.g., *Examiner of San Francisco, Chicago Tribune,* the *Philadelphia Inquirer, Star of Montreal, Star Telegram of Fort Worth, Los Angeles Herald Examiner,* and *Register of New Haven*) had no record of printing any article related to ginseng.

Modern German, French, and Anglo-American standard textbooks of medicinal chemistry, pharmacology, and therapeutics, and other medical reference books do not contain ginseng in their index. Textbooks of pharmacognosy are the only professional books that provide information on ginseng. One pharmacognosy book says: "Ginseng is a stimulant and a stomachic. It is a favorite remedy in Chinese medicine."[7] Another pharmacognosy book states: "Ginseng contains glycoside called Panaquilon, panax sapoginol, volatile oil, a physosterin, mucilage, a sugar, starch, etc. Ginseng is used by the laity as a stimulant and aromatic bitter. The Chinese employ it as an aphrodisiac and heart tonic but without scientific identification."[8] A third pharmacognosy book, which is the newest one, refers to ginseng: "Ginseng is a favorite remedy in Chinese medicine and is considered to have tonic, stimulant, diuretic, and carminative properties. It reportedly reduces the blood sugar concentration and acts favorably on metabolism, the central nervous system, and on endocrine secretion. It is employed in the Orient in the treatment of anemia, diabetes, insomnia, neurasthenia, gastritis, and especially, sexual impotence."[9]

American ginseng remains on the official *United States Pharmacopeia* (USP)[10] list of acceptable herb drugs from 1842 to 1882 but only as a supplemental drug for stomachic and stimulant uses. During the first half of the twentieth century, ginseng was officially listed as a demulcent in the *United States Dispensatory* (USD)[11] and *National Formulary* (NF),[12] as well as USP. However, ginseng was dismissed as a therapeutically useless drug and was deleted from the official compendia in 1950; that was the end of American ginseng, and since then, nobody wants to talk about it. For about 25 years, ginseng has been unknown to the majority of the American public.

GAP OF SCIENTIFIC INFORMATION

There has been a serious problem finding scientific information on ginseng in the Western countries, since nearly all ginseng research work has been conducted in the Orient and in the Soviet Union and published in their native languages of Chinese, Korean, Japanese, and Russian, in difficult-to-obtain or obscure journals. No library in the Western countries, including the Library of Congress and the National Library of Medicine, has a complete collection of foreign journals.

The *Index Medicus*[13] does not list as many ginseng articles as did the *Chemical Abstract*.[14] I made a complete survey and found about 300 pieces of ginseng information listed in the entire *Chemical Abstract* starting with the first volume of 1907 up to volume 84 of the current year. By no means is it a complete collection, for it is impossible for the Chemical Abstract Service to collect and translate all of the foreign papers of every discipline. The *Chemical Abstract* is not easily accessible to the American public.

American professional journals have not published any ginseng articles in the last 25 years (before 1970), since few studies on ginseng were conducted in the United States. Questions often are raised nowadays by many customers to their pharmacists, by patients, many retired people, and some business-minded ginseng enthusiasts who want to grow ginseng or formulate ginseng for products, yet not one single booklet that is currently available tells the modern scientific findings about ginseng. A letter from the Agricultural Research Center of the U.S. Department of Agriculture written to me said, "Although the Department does not conduct research studies on ginseng every year we receive thousands of request for information on its culture, marketing and uses."[15] Another correspondent, also from the Agricultural Research Service Center at Beltsville, Maryland, wrote to me: "With the increased interest in acupuncture, I note an increased interest in Oriental medicinal plants. I get more inquiries about ginseng than about any other medicinal plant."[16] The American public is desperately searching for new evidence about ginseng.

MODERN SCIENTIFIC RESEARCH AND
PUBLICATIONS ON GINSENG

The earliest chemical and pharmacological studies on ginseng started in the mid-nineteenth century. As a matter of fact, research was started right here in the United States with the American ginseng root collected in Canada. The first chemical report was published in 1854.[17] At that time it was extremely difficult to obtain any large quantity of authentic specimens of American wild ginseng, and also due to the extravagant prices, studies were limited, and little fruitful results were made before 1900.

After the turn of the twentieth century, most of the preliminary studies were conducted by the Japanese. From 1900 to 1935, ginseng studies became one of the most active projects at many pharmaceutical colleges and medical centers. As a result of the annex of Korea by the Japanese military and later their occupation of Manchuria, the Japanese then possessed all the valuable roots, as they had desperately desired for centuries. Within 35 years' time, about 50 studies were conducted in the areas of chemical, biochemical, pharmacological, and even clinical trials on ginseng extracts—not only ginseng from Manchuria and Korea, but also specimens of American ginseng. Although their results did not reach a clear conclusion that ginseng is useful for the treatment of certain ailments, their results stimulated the interest of scientists in Western countries including the Russian Federation.

From 1936 to 1945, ginseng research was interrupted as a result of the Sino-Japanese War (1937–1945) and World War II (1939–1945). Few ginseng scientific papers are seen in the literature published between 1936 and 1950.

Being the next-door neighbor to China, the Russians, following the track of the Japanese, showed tremendous interest in Chinese ginseng. As early as in 1859, a scientific expedition was organized in St. Petersburg (now Leningrad) to explore the Manchu territory and the frontier of Korea and examine especially the ginseng and other Chinese plantations with a view of establishing plantations on Russian soil for the purpose of trade with China.[18]

Prior to the Second World War, limited studies on ginseng were conducted at the Institute of Experimental Medicine of the Russian Federation. During 1945 and 1948, the Red Army invaded Korea and Manchuria; thus, the ginseng-rich lands were then transferred from the Japanese to the Russians. Supposedly the Red Army then grabbed the entire supply of ginseng valued at about $120 million, and shipped it home for studies and consumption. A special institution called the Institute of Biologically Active Substances, Siberian Department of the Academy of Sciences of the USSR, was established in Vladivostok in 1949. The function of the newly established organization was to conduct research on medicinally useful herbs found in China and Siberia, particularly ginseng. The Vladivostok center coordinates all related pharmacologic, clinical, and chemical investigations from Vladivostok, Novosibirsk, Khabarovsk, Leningrad, Tomak, Moscow, and many other cities of the Russian Federation. In 1965, this institute moved into an excellently equipped new laboratory building erected on the picturesque shore of the Armur Ray. A special ginseng committee uniting the efforts of all scientists and clinicians engaged in this field has been working since 1949. Up to 1965, the committee held 23 sessions in which not only Soviet but foreign scientists participated as well. The committee has issued seven volumes of works and a two-volume collection of the minutes of the committee sessions.[19] Through extensive research, the Soviet scientists, in collaboration with the Japanese, have established the nature, structure, and medicinal properties of the active principles of ginseng.

In South Korea, as well as in China, ginseng research work, mostly confined to the pharmacologic and physiologic aspects, was started after 1956. New methods of growing and of controlling typical ginseng diseases were also explored. Little chemical work, however, was published, as is true today.

In 1960, a new era of ginseng research was started. As a result of the availability of modern instrumentation and techniques, in addition to an ardent interest in ginseng research in Asia and some parts of Europe, significant accomplishments have been made.

In the last 15 years, at least 160 chemical papers have been published. Elyakov and his associates at Vladivostok and Shibata and his coworkers at the Faculty of Pharmaceutical Sciences, University of Tokyo, have successfully isolated and identified the ginseng saponins from *Panax ginseng* root as well as from other species. Scientists who have similarly contributed significantly toward the understanding of ginseng chemistry are Horhammer of the University of Munich; Kondo of the School of Pharmaceutical Sciences, Showa University, who established the structures of chemical constituents of Japanese ginseng; and Staba of the College of Pharmacy, University of Minnesota, whom with his coworkers has been studying extensively the American ginseng plant and the mechanism of biosynthesis of ginseng saponins.

In the biological studies of ginseng from 1930 to 1960 by Japanese workers; the Soviet scientists at Vladivostok; and by Bulgarian, Chinese, and South Korean scholars, they all used crude ginseng extracts, either aqueous or alcoholic, as the testing materials. Researchers began to evaluate the activities of ginseng by using the purified saponins as testing material, or the different fractions of extraction since 1970. This is, no doubt, the beginning of a new milestone in ginseng research. Thus, the biological data on ginseng's activities, using chemically purified material and identified specimens, are much more scientifically meaningful and valid. Much remains to be done, however, in the study of the biomedical and biochemical actions of ginseng, and its clinical trials must still be conducted.

GINSENG SYMPOSIUM

The first international ginseng symposium, sponsored by the Central Research Institute of the Korean Office of Monopoly, convened in Seoul in September 1974.[20] Some 500 well-known scientists, pressmen, businessmen, and government officials from Belgium, Britain, Japan, Singapore, Sweden, South Korea, Switzerland, Taiwan, Thailand, the Netherlands, Turkey, the United States, and West Germany attended the meeting. The ginseng scientists and ginseng enthusiasts sat together to discuss the different aspects of ginseng, its pharmacology, composition, clinical aspects, and modern methods of cultivation.

The meeting opened with a keynote speech by the president of the symposium, Choi Yoon-kuk, director of the Central Research Institute of the Office of Monopoly. "The symposium is designed to provide an opportunity for international exchange of academic and scientific information concerning ginseng," Choi said. Lee Chang-suk, the vice minister of the Ministry of Science and Technology, echoed the same theme with a reminder that "Research on the properties of ginseng is proceeding both in the Occident and the Orient."

Regarding the pharmacological properties of ginseng, research papers were given by, respectively, K. Takagi of the University of Tokyo; Hong Sa-ak of the College of Medicine, Seoul National University; D. Ivan Popvov of the Renaissance Revitalization Center at Nassau, Bahamas; Finn Sandberg of the University of Uppsala, Sweden; Karl Reuckert, managing director of Pharmaton, Ltd. in Switzerland; and Lee Kwang-soo of the State University of New York. In addition, a review of information on ginseng pharmacological work was presented by Karlfried Karzel of the University of Bonn.

In the areas of chemical composition and biochemical effects of ginseng, papers were presented by L. Horhammer of Munchen University, S. Shibata of Tokyo University, S. Hiai of Toyama University, M. Yamamoto of Osaka University, and John Staba of the University of Minnesota.

On the cultivation of ginseng, Lee Choon-young of Suwon Agriculture College reported his study of the chemical quality of the nursery and soil of ginseng cultivation. Yeng Young-tu of the Taiwan Agricultural Institute reported the effect of various mulching cultivations on ginseng root and quality. Changyawl Harn of the Korea Atomic Energy Research Institute reported his study of the systematic cultivation of ginseng.

The second international ginseng symposium, sponsored by Pharmaton Ltd. and the World Health Organization, convened April 9 to 12, 1975, in Lugano Switzerland.[21] Similarly, the objective of the meeting was to discuss scientific research and exchange ideas on ginseng.

In a 122-year history of ginseng's research, about 750 scientific articles have been published altogether by a few hundred scientists in 12 countries. Among these papers, 70% are biological studies, 20% are chemical, and the rest are botanical and reviews. Interests include the broad clinical trials and the chemical synthesis of the ginseng saponin structure out of economical sources. Finding ginseng substitutes is another project scientists are seriously undertaking with keen interest.

REFERENCES

1. Jones, B., *The New York Times*, March 14, 1971.
2. *The New York Times*, April 25, 1972.
3. King, W., *The New York Times*, October 11, 1975.
4. Geczl, M. L., *The Wall Street Journal*, September 9, 1975.
5. *Milwaukee Journal Sentinel*, April 8, 1973.
6. Baumgart, R. A., *Milwaukee Journal Sentinel*, December 13, 1974.
7. Gathercoal, E. N. and E. H. Wirth, *Pharmacognosy*, Lea and Febiger, Philadelphia, 1936.
8. Youngken, H. W., *Textbook of Pharmacognosy*, The Blakiston Company, Philadelphia, 1950, pp. 607–609.

9. Claus, E. P., U. E. Tyler, and L. R. Brady, *Pharmacognosy*, Lea and Febiger, Philadelphia, 1970, pp. 113–114.

10. *United States Pharmacopeia*, Mack Publishing Company, Easton, Pennsylvania.

11. *United States Dispensatory*, Mack Publishing Company, Easton, Pennsylvania.

12. *National Formulary*, J. P. Lippincott Company, Philadelphia.

13. *Index Medicus*, U.S. Government Printing Office, Washington, DC.

14. *Chemical Abstract*, Chemical Abstract Service, Columbus, OH.

15. Webb, R. G. to J. P. Hou, Correspondence, April 18, 1975.

16. Duke, J. A. to J. P. Hou, Correspondence, June 2, 1975.

17. Garriques, S., *Ann. Chem. Pham.*, 90: 231, 1854.

18. Jackson, J. R., *Pharm. J.*, 86, 1875.

19. Brekhman, I. I., *Ind. J. Pub. Health*, 9: 148, 1965.

20. "Proceedings of International Ginseng Symposium," *Central Research Institute, Office of Monopoly*, Seoul, Korea, 1974.

21. *Scientific Documentation of Panax Ginseng C. A. M.*, Pharmaton Ltd., Lugano-Bioggio, Switzerland, 1975.

12 Less Known Benefits of American Ginseng

American ginseng, scientific name *Panax quinquefolius*, is a close relative of Chinese ginseng (*Panax ginseng*), and belongs to the Araliaceae family, which is the same as Siberian ginseng (*Eleutherococcus senticosus*). It is a perennial herb, distinguished by its dark-green leaves and clusters of red berries, that grows wild in eastern North America. The root of the plant is used medically, particularly in China where traditional Chinese medicine places a high value on it.

Of the traditional ginsengs, American ginseng is probably the least used and researched variety. Americans have never been large consumers of American ginseng. In the past, American ginseng was an uncommon folk remedy used as mild stimulant, tonic, and digestive aid. Most of this herb was exported to China, where most ginseng is consumed. American ginseng is considered a less potent member of the ginseng family, but it is a highly prized tonic and herbal remedy.

American ginseng was used medicinally by many Native American tribes as a health stimulant and sexual tonic and for various health problems including headaches, female infertility, digestive problems, fever, and earache. American ginseng was introduced by Native Americans to European settlers in North America in the early 1700s. A French Jesuit priest named Jartoux had traveled through China and was convinced of the medicinal powers of Korean ginseng. In 1714, he published a paper in Britain about Korean ginseng and its healing powers, and he theorized that the plant may grow wild in the favorable climate of North America. Another Jesuit missionary in Canada, Joseph Lafitau (1681–1746), read the article and began searching the woods near his dwelling. Lafitau found American ginseng plants, which bear a close resemblance to their Asian cousins, and sent samples of them to China. A thriving trade of American ginseng began around 1718: it was sent to the Orient after being gathered by Native Americans, French fur traders, and early frontiersmen including Daniel Boone.

The majority of American ginseng on the market is now cultivated, although it is a sensitive plant and difficult to farm. In the United States, Wisconsin grows 80% of the American ginseng crop. Canada grows more American ginseng than any country and is second only to China in total ginseng production.

SCIENTIFIC RESEARCH

The majority of research performed on ginseng has been done on the Chinese and Siberian varieties. Clinical and chemical research on American ginseng is yet to be done. One reason for this is the American medical establishment's skepticism of herbal remedies.

American ginseng is classified as an adaptogen, which is a substance that helps the body adapt to stress and improves immune response. Adaptogens must also be nontoxic and cause no major physiological changes or side effects. American ginseng root has an array of complex chemicals, and scientists have determined that the active ingredients are saponin triterpenoid glycosides, or chemicals commonly called *ginsenocides*. American ginseng contains nearly 30 ginsenocides. However, American ginseng has been found to contain higher levels of ginsenocide Rb1, which has a sedative effect on the central nervous system, than other species of ginseng. Thus, scientific research has been consistent with Chinese herbalists' claims that American ginseng is less stimulating than Chinese ginseng. The research implies that American ginseng can provide the strengthening and immune-enhancing effects of other ginseng without overstimulation to those people with high levels of stress and mental stimulation. American ginseng may be the best ginseng for Americans, whose fast-paced and energetic lifestyles may call for more calming and balancing herbs.

GENERAL USE

American ginseng can be used by those people who seek the adaptogenic effects (toning, strengthening, and immune-enhancing effects) of *Panax ginseng* without the highly stimulating aspects. Chinese herbalists consider American ginseng to be a cooling herb, so it can be used as a tonic and immune strengthener for people who are overstressed or suffer from hot conditions like high blood pressure, excess nervous energy, or ulcers. American ginseng, according to Chinese herbalists, is more suitable and balancing for women and children than Panax ginseng, and is more applicable for the elderly who wish to avoid stimulants. American ginseng is also used in Chinese medicine for chronic fevers; to aid in the recovery of infectious diseases; for strengthening the lungs in cases of tuberculosis, bronchitis and asthma; and for the loss of voice associated with respiratory disorders.[1-3]

Of the traditional ginsengs, American ginseng is probably the least used and researched variety. Americans have never been large consumers of American ginseng. In the past, American ginseng was an uncommon folk remedy used as a mild stimulant, tonic, and digestive aid. Most of this herb was exported to China, where most ginseng is consumed. American ginseng is considered a less potent member of the ginseng family, but it is a highly prized tonic and herbal remedy.

REFERENCES

1. https://www.encyclopedia.com/medicine/divisions-diagnostics-and-procedures/medicine, March 14, 2018.
2. https://www.encyclopedia.com/medicine/divisions-diagnostics-and-procedures/medicine, February 28, 2018.
3. American ginseng facts, information, pictures|Encyclopedia.com, articles about American ginseng.

Section V

Composition of Ginseng

13 Chemical Constituents of Ginseng Plants

The chemical constituents of the following are discussed in this chapter:

1. Asian ginseng (*Panax ginseng*)
2. American ginseng (*Panax quinquefolius*)
3. Japanese ginseng (*Panax japonicus*)
4. Siberian ginseng (*Eleutherococcus senticosus*)
5. *Panax notoginseng*

CHEMICAL CONSTITUENTS OF ASIAN GINSENG

The Bible says that for every disease there is an herb in the plant kingdom. The plant kingdom provides a tremendous reservoir of natural crude drugs. In the past thousands of years, people all over the world, particularly the Chinese, have consumed considerable quantities of botanical drugs, more frequently called herbal drugs. Herbals remain a primary source of many clinically useful remedies.

The virtue of the herbal remedies is that most of these natural products have proven to be medicinally effective, mild, and relatively safe. *Ma Huang*, described as a remedy for coughs and as a cardiac stimulant, is one of the oldest Chinese herbals. Liquorice root has been used in medicine for treating stomach complaints and as a general sweetener in China as well as in other countries for many centuries. The snakeroot (*Rauwalfia cerpentina*) has been used by the Indians as a tranquilizer for more than 400 decades.

How do ancient herbals become modern medicine? It is true that a great number of modern remedies used in the United States are from a natural source. You must have used or at least heard of ephedra, ergot, digitalis, penicillin, codeine, morphine, cocaine, quinine, atropine, and hundreds of others derived from nature. To obtain therapeutically useful principles from the crude herbs, as a rule, the medicinally useful or biologically active principles are isolated by suitable extractions, separations, and purifications. The important principles contained in the crude drugs are called *constituents*. Some of the constituents may be biologically active, while others may be inert and thus medicinally valueless. The most commonly encountered principles such as cellulose, lignin, suberin, cutin, starch, and coloring matter are inert materials, whereas glycosides, alkaloids, vitamins, enzymes, volatile oils, fixed oils, and acids are important biologically active constituents. In the old days, crude water extractions or decoctions of drugs were frequently used in the treatment of diseases. In modern times, however, crude extracts are usually further processed to obtain purified principles. In order to find out how good the herbal drug, its extract, or its purified materials are, pharmacological and toxicological studies are carried out with laboratory animals. With promising animal evidence, preliminary clinical trials in humans may be conducted. Efficacy and long-term toxicity studies are the last steps in developing a new drug. To develop a new chemical drug, drug companies usually spend an average of 5–10 years and more than $12,000,000 from its synthesis to marketing. With the development of drugs out of natural sources, the cost may be even higher.

We recall that soon after isolation of reserpine from the snakeroot, it became the most important tranquilizer in the treatment of hypertension. Another example is aspirin. After the discovery of salicin in the bark of willow trees, it has been developed as the most versatile and most widely prescribed drug in history. Even in modern times, since the discovery of penicillin in 1929, thousands of new penicillins have been prepared, and a great many have been used clinically.

Among the five botanically identified *Panax* species (*Panax ginseng, P. quinquefolium, P. trifolium, P. japonicum, P. pseudoginseng*), all except *P. trifolium* have been used as crude drugs in the Orient. Although Siberian ginseng (*Eleutherococcus senticosus*) does not belong to the genus *Panax*, it possesses similar medicinal activities. All of these plants are botanically related—that is, they all belong to the family Araliaceae. These plants do not, however, contain the same active constituents, and they do not possess the same medicinal values. The kinds of chemical principles contained in the various *Panax* or ginseng plants are discussed in this chapter.

CHEMICAL RESEARCH ON *PANAX GINSENG*

In the course of chemical studies on ginseng, chemists have attempted to isolate and identify ginseng's active principles, and thus verify the usefulness of ginseng.

Rafinesque seemed to be the first who reported his findings of a camphor-like substance, *Panacene*, from an extract of American ginseng root in 1854.[1] Almost simultaneously, Garriques reported his findings that from a crude extract of American ginseng root, he obtained a yellowish glycoside called *Panaquilon*.[1,2] Panaquilon is soluble in water and alcohol but insoluble in ether. Its sweet taste resembles that of licorice root. After acid hydrolysis, panaquilon gives a water-insoluble substance called *Panacon*. Later in 1889, a Soviet worker, Davydow, reported that he similarly obtained a ginseng glycoside from extracts of Manchurian ginseng roots, and its properties were similar to those of panaguilon.[3]

Pioneer studies were also conducted by quite a number of Japanese chemists during the 1900s. Inoue isolated a substance from an extract of Japanese ginseng (*chikusetsu ninjin*).[4] This substance was shown to be a saponin glycoside. It has some hemolytic (breaking down of red blood cells) properties. In 1905, *Fujitani* isolated a relatively pure saponin from the extracts of Korean and Japanese ginseng roots.[5] It was a white crystalline powder and was similarly called *panaquilon*. In the following year, Asahina and Taguchi reported that from an alcoholic extract of ginseng root, a noncrystalline substance was obtained.[6]

More detailed chemical studies on ginseng were conducted by Kondo and his associates. The powdered ginseng roots were extracted with water, methanol, and ether, respectively. A mucilaginous substance and some inorganic substances were found in the aqueous extract; a mixture containing cane sugar, nitrogenous substances, and a saponin glycoside was obtained from the methanol extract; and an oily substance, brown in color with typical ginseng fragrance, was obtained from the ether extract. The ginseng oil was further fractionated with steam distillation into two portions—that is, a light yellowish volatile oil, and a brown, nonvolatile portion. The volatile oil contained panacene, while the nonvolatile fraction was shown to contain phytosterol and fatty acids. Upon hydrolysis of the saponin glycoside with 7% alcoholic-hydrochloric acid, two new substances, panax-sapogenol and panax-sapogenol-amorphous (noncrystalline), were obtained. Further studies showed that the ginseng saponin glycoside consisted of 1 mole of panax-sapogenol, 2 moles of glucose, and 1 mole of pentose, and the molecular weight of the saponin glycoside was 876. From quite a number of authentic samples, it was found that the cultivated Korean and Japanese ginseng gave similar chemical components.[7–9] Abe and Saito[10] and Yonekawa[11] together isolated a relatively pure ginseng saponin glycoside, ginsenin, from Korean ginseng root. Pharmacologic studies were conducted with ginsenin. In 1930, Kotake similarly isolated a ginseng saponin that was called panaxin, but this substance did not show hemolytic properties. After hydrolysis of panaxin with 50%sulfuric acid in methanol, a new substance, α-panaxin, was obtained. It was the sapogenin (aglycone) of ginseng extract.[12] Then α-panaxin was further degraded with fuming hydrochloric acid, giving glucose and a halogenated compound called aglucone. Sakai had more interest in the fragrant ginseng oil. From an ether extract of ginseng roots, a considerable amount of sweet-smelling and biologically active ginseng oil was obtained. The most active component of the oils that represents the activities of the whole ginseng was the volatile oil portion called panacene. The ether extract also contained a number of other ingredients. High molecular weight unsaturated fatty acids, called panax acids, and esters of fatty acids have been isolated.[13–16]

In the last 16 years, great advancement has been made in isolation, characterization, and identification of the principles of ginseng, and ginseng saponin chemistry, first in history, has become known to the scientific world.

By using different solvent systems, it was found that some of the ginseng constituents are only extractable with water or alcohols, while others are primarily organic solvent or oily soluble. The crude extracts have been successfully separated with the aid of thin-layer chromatography (TLC) and column chromatography techniques. TLC has been a very useful tool that separates ginseng saponins of similar structures. Infrared (IR), nuclear magnetic resonance (NMR), and mass spectroscopy have provided a powerful means for elucidating the structures of the useful principles contained in the mysterious root.

Based on the chemical studies of ginseng, we have learned that *Panax ginseng* root contains a number of biologically active and medicinally useful components.[17–19] According to their chemical nature, these numerous components, so far identified, can be classified into nine groups:

- Ginseng saponins (saponin glycosides)
- Ginseng oil and phytosterol
- Sugars and carbohydrates
- Organic acids

- Nonprotein nitrogenous substances
- Amino acids and peptides
- Vitamins
- Minerals and trace elements
- Unknown enzymes

It has been confirmed pharmacologically that the chief biologically active components in ginseng and other *Panax* species appear to be ginseng saponins. Ginseng roots contain a substantial amount of ginseng saponins; accordingly, the saponins contribute most, if not all, of the important pharmacologic activities of ginseng extract. The chemical properties of these important ginseng principles are discussed further in the following sections.

GINSENG SAPONINS

Saponins, a sweet-bitter material, usually exist in plants in the form of glycosides known as *saponin glycosides*. Ginseng, sarsaparilla, glycyrrhiza, etc., contain a considerable amount of saponin glycosides. Saponin glycosides are giant molecules that are extractable from the plant materials by hot water or alcohols. Saponins have particular chemical properties, and the most notable characteristics are as follows: they form colloidal solutions in water that foam upon shaking (frothing); they have a bitter taste; they have sternutatory and irritative properties on mucous membranes; and they have hemolytic action against red blood cells (Table 13.1).[20,21]

A number of ginseng saponin glycosides have been isolated from the methanol extracts of ginseng roots and identified by the Soviet workers at Vladivostok and Moscow. These saponins are called *panaxosides*.[22,23] In 1962 Elyakov et al. reported that they were successful in isolating saponin glycosides, panaxosides A and B.[24] Two years later, four additional saponins, panaxosides C, D, E, and F, were isolated.[25] Based on composition, these six ginseng saponin glycosides belong to two groups. Panaxosides A, B, and C belong to one group, and they yield a common nonsugar substance (aglycone) called *panaxatriol*, while panaxosides D, E, and F belong to

TABLE 13.1

Composition and Physical Properties of Ginseng Saponin Glycosides Panaxosides Isolated from *P. ginseng* C. A. Meyer

Panaxoside	Melting Point (°C)	Molecular Weight	Composition Genin (sapogenin)	+ sugar (mol)
A	176–178	—	Panaxatriol	Glucose (3)
B	182–185	—	Panaxatriol	Glucose (2) + Rhamnose (1)
C	185–187	1031–1064	Panaxatriol	Glucose (3) + Rhamnose (1)
D	157–160	1178	Panaxadiol	Glucose (4)
E	185–187	1222–1230	Panaxadiol	Glucose (4) + Arabinose (1)
F	185–187	1388–1424	Panaxadiol	Glucose (6)

another group and yield panaxadiol after acid hydrolysis. None of the six panaxosides carries identical sugars. Panaxoside A consists of three glucoses; panaxoside B, two glocoses and one rhamnose; panaxoside C, three glucoses and one rhamnose; panaxoside D, four glocoses; panaxoside E, four glocoses and one arabinose; panaxoside F, six glucoses.[25,26] The composition and some of the physical properties of panaxosides are listed in *J. Pharm. Soc., Japan* 13.1. In 1974, Andreev et al. were able to isolate 14 spots on the TLC plates, each spot representing one different ginseng glycoside or panaxoside.[27] In cooperation with the Japanese workers, the structures and the properties of these panaxosides have been identified.

In 1962, Funjita, Itokawa, and Shibata of the University of Tokyo were able to isolate two types of compounds, saponin glycosides and sapogenin (aglycones), from the methanol extracts of three different species of ginseng roots (ginseng, Japanese ginseng, and American ginseng).[28] The ginseng saponin glycosides isolated by the Japanese workers were called ginsenosides, which is quite different from the Soviet nomenclature (panaxosides). *Sapogenins* (the nonsugar portion of saponin glycoside) are normally obtained when the saponin glycosides are treated with hot hydrochloric acid (HCl) in methanol. Soon it was learned that the sapogenins thus obtained were not true sapogenins but actually the acid hydrolyzed products of ginseng saponins. Soon the structure of the acid hydrolyzed saponin was identified as panaxadiol.[29-31] Immediately Shibata's group was successful in confirming that the structure of panaxadiol is a tetracyclic, triterpene of dammarane structure.[31,32]

The genuine sapogenin of ginseng was eventually obtained when the ginseng extracts or the saponins were hydrolyzed under mild conditions (0.7% sulfuric acid) in methanol solution. The true sapogenin, thus obtained, was called *prosapogenin*,[33] and the chemical structure of it was identified as protopanaxadiol.[34]

After a series of comparative studies, it is confirmed that panaquilon (originally isolated by Garriques), panaxin (isolated by Kotake), and ginsenin (isolated by Yonekawa) are the same saponin. Panacon (isolated by Garriques), α-panaxin (isolated by Kotake in 1930), and a compound that melted at temperatures greater than 270°C obtained by Asahina et al. are identical to prosapogenin, the "genuine aglycone of ginseng," which melts at 330°C.[28]

In 1965, more than 10 neutral ginseng saponins were isolated by the TLC technique, and they were designated as ginsenoside Rx (where $x =$ a, b, c, d, e, f, g1, g2, g3, and h). Hydrolysis of ginsenosides Rb and Rc, panaxadiol, was obtained. Similarly, hydrolysis of ginsenoside Rg_1, however, a new sapogenin, panaxatriol, was isolated. All physicochemical data indicate that panaxatriol is a homolog of panaxadiol and carries one more OH group than panaxadiol.[35]

Shibata et al. in 1974 successfully isolated 13 ginsenosides from ginseng root extract. They are Ro, Ra, Rb_1, Rb_2, Rc, Rd, Re, Rf, Rg_1, Rg_2, Rg_3, Rh_1, and Rh_2. The ginsenosides from Ro to Rh correlate the increased Rf value of the TLC. Ginsenoside Ro is the least polar, having the lowest Rf value. From the TLC plates, ginsenosides Rb_1 and Rb_2 (Rb group) and Rg_1, Rg_2, and Rg_3 (Rg group) are noted to be the main components of the saponins.[36]

The detailed compositions or structures of these saponins were not known until 1968, when Iida, Tanaka, and Shibata reported that ginsenoside Rg_1 is composed of protopanaxatriol and two glucose molecules. At present, 9 of the total 13 ginsenosides

TABLE 13.2
Physical Properties of Ginsenosides

Ginsenosides	Physical Properties	Melting Point (°C)	Formula	IR (K Br) cm¹
Ro	Colorless needles (MeOH)	239–241	$C_{48}H_{76}O_{19}$	3400 (OH) 1740 (COOR) 1728 (COOH)
Rb_1	White powder EtOH-BuOH	197–198	$C_{54}H_{92}O_{23}$	3400 (OH) 1620 (C = C)
Rb_2	White powder EtOH-BuOH	200–203	$C_{55}H_{90}O_{22}$	3400 (OH) 1620 (C = C)
Rc	White powder (EtOH-BuOH)	199–201	$C_{53}H_{90}O_{22}$	3400 (OH) 1620 (C = C)
Rd	White powder (EtOH-AcOEt)	206–209	$C_{48}H_{82}O_{18}$	3400 (OH) 1620 (C = C)
Re	Colorless needles (50% EtOH)	201–203	$C_{48}H_{82}O_{18}$	3380 (OH) 1620 (C = C)
Rf	White powder (aceton)	197–198	$C_{42}H_{72}O_{14}$	34,380 (OH) 1620 (C = C)
Rg_1	Colorless powder	194–196	$C_{42}H_{72}O_{14}$	No OH
Rg_2	Colorless powder	187–189	$C_{42}H_{72}O_{13}$	3400 (OH) 1620 (C = C)

have been characterized, and their structures have been established.[36–38] The physical form, melting point, chemical formula, and characteristic IR of the nine ginsenosides are listed in Table 13.2.

According to the nonsugar portion of the molecule, these ginsenosides can be divided into three groups. Ginsenoside Ro gives oleanolic acid (aglycone), while ginsenosides Rb, Rc, and Rd give panaxadiol, and ginsenosides Re, Rf, and Rg give panaxatriol after acid hydrolysis. Ginsenoside Ro is quite a different compound in comparison with other ginseng saponins, but it is identical to chikusetsu saponin V, a saponin isolated from Japanese ginseng.

Each of these ginsenosides carries a somewhat different sugar molecule, though they all carry monosaccharides. Table 13.3 shows the composition of nine ginsenosides of panax ginseng.

The chemical structure of panaxadiol (structure I) was elucidated by Shibata et al. in 1963. In early reports the compounds sapoginol isolated by Kondo et al. in 1915–1920, panaxol isolated by Wagner-Jauegg and Roth in 1963,[39] and ginsengenin isolated by Lin[40] are shown to be identical to panaxadiol.[37]

The structure of protopanaxadiol (structure II) has been confirmed to be an open-chain compound, having a free alcohol group, and an end vinyl group. Drastic acid hydrolysis of protopanaxadiol forms panaxadiol.

The structure of panaxatriol (structure III), similarly, is an acid-hydrolyzed product during isolation. The genuine alycone would be protopanaxatriol (structure IV).

TABLE 13.3
Composition of Ginsenosides

Ginsenosides	Prosapogenin (aglycone)	+ Sugar component (mole)
Rbo	Oleanolic acid	Glucose (2), glucuronic acid
Rb_1	Protopanaxadiol	Glucose (4)
Rb_2	Protopanaxadiol	Glucose (3), arabinose (1)
Rc	Protopanaxadiol	Glucose (3), arabinose (1)
Rd	Protopanaxadiol	Glucose (3)
Re	Protopanaxatriol	Glucose (2), rhamnose (1)
Rf	Protopanaxatriol	Glucose (2)
Rg_1	Protopanaxatriol	Glucose (2)
Rg_2	Protopanaxatriol	Glucose (1), rhamnose (1)

Oleanolic acid (structure V) is a pentacyclic, oleanane-type triterpene compound that was initially isolated and identified by Horhammer et al.[41]

The structures of several ginseng saponins, ginsenoside Rb_1 (structure VI), ginsenoside Rg_1 (structure VII), and ginsenoside Ro (structure VIII) were eventually established and confirmed (Figure 13.1). For example, the structure of Rg_1 has been established as 6,20-di-O-β-glucosyl-20-S-protopanaxatriol (Figure 13.2).

Also confirmed was that panaxoside A, a ginseng saponin isolated by the Soviet chemist Elyakov et al., was identical to ginsenoside Rg_1.[24]

OTHER COMPONENTS OF GINSENG ROOT

The most important components, other than saponins, from the extraction of ginseng root are ginseng oil, phytosterol, carbohydrates, acids, nitrogenous substances, vitamins, minerals, enzymes, and ferments (Figure 13.3).[17–19]

GINSENG OIL AND PHYTOSTEROL

Ginseng oil contains low boiling or volatile, and high boiling or nonvolatile, fractions. The low boiling fraction (boils from 70°C to 100°C) contains panacene and c-elemene. Panacene gives the characteristic fragrance of ginseng. The high boiling fraction (boils from 120°C to 150°C) contains panaxynol.[42,43] Panaxynol, a yellow oil, is a straight-chain, unsaturated alcohol with 17 carbon atoms (Figure 13.4).[44,45]

β-Sitosterol, isolated from the nonvolatile fraction of ginseng oil by Horhammer, is the most common phytosterol in plants.[41] It has been used as a competitive cholesterol inhibitor. Analysis of the phytosterols extracted from Chinese and Korean ginseng revealed that these phytosterols are distinctive for each species.[46] The phytosterol from Chinese ginseng is called stigmasgterol,[40] and that from Korean ginseng is called β-sitosterol.[46]

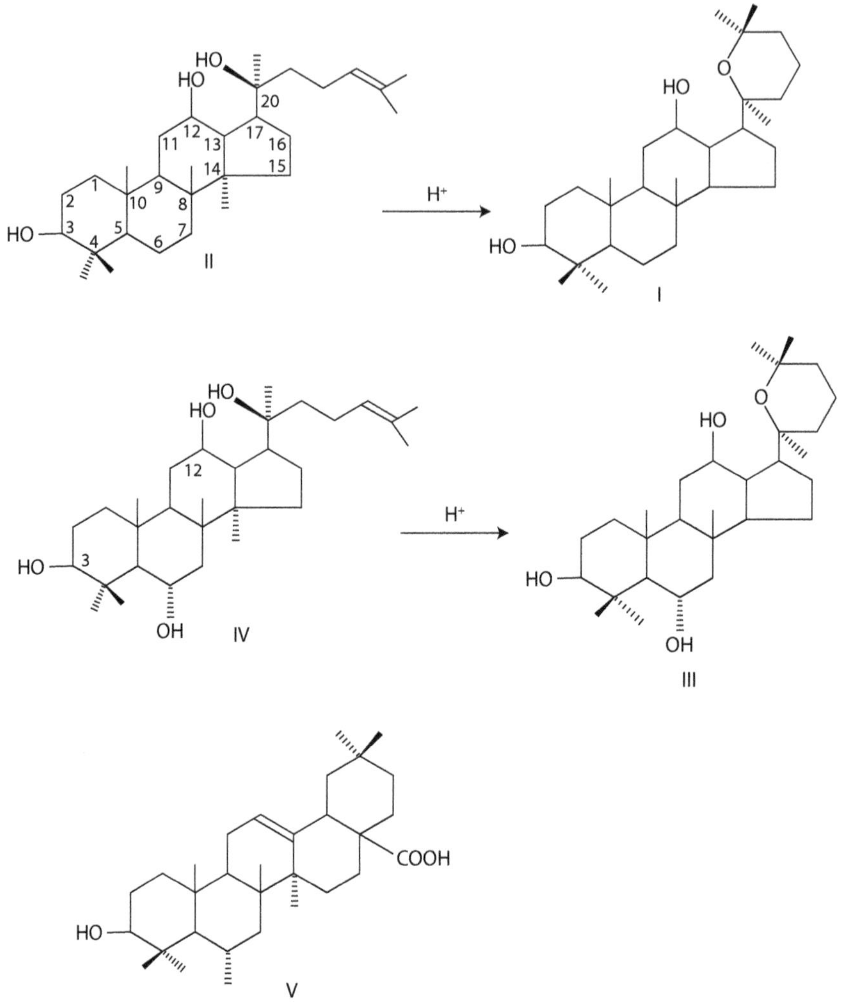

FIGURE 13.1   Structures of ginseng saponins.

Sugars and Carbohydrates

The aqueous extract of ginseng root contains many different types of sugar or saccharide.[47–49] The most common single-sugar compounds (monosaccharides) are glucose and fructose. Maltose and sucrose are the most common double-sugar compounds (disaccharides). Multisugar compounds such as trisaccharides, tetrasaccharides, and oligosaccharides are also found in ginseng root.

A crude polysaccharide, which was identified as ginseng pectin, was isolated from a boiling-water extract of ginseng root.[50–52] Pectin is a high molecular weight polysaccharide. It is normally found in plants and fruits. It has been used widely in food preparations such as jellies and jams, as a plasma expander, and medicinally as an antidiarrheal agent.

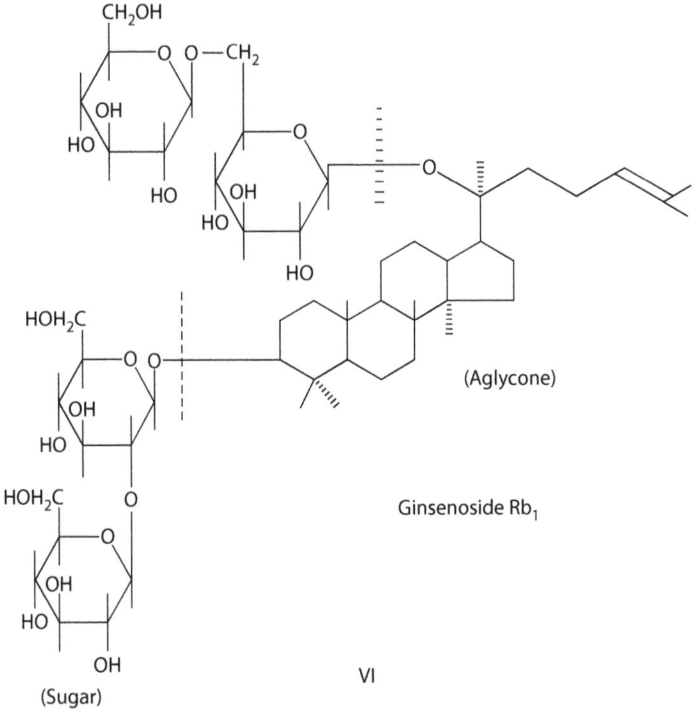

VI

FIGURE 13.2 Ginseng saponin.

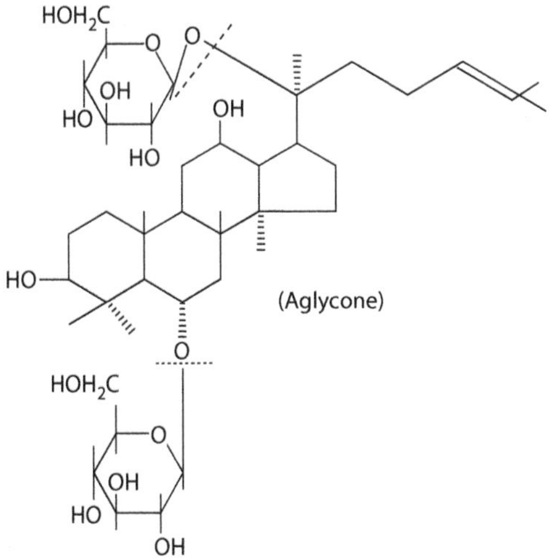

VII

FIGURE 13.3 Ginsenoside Rg_1.

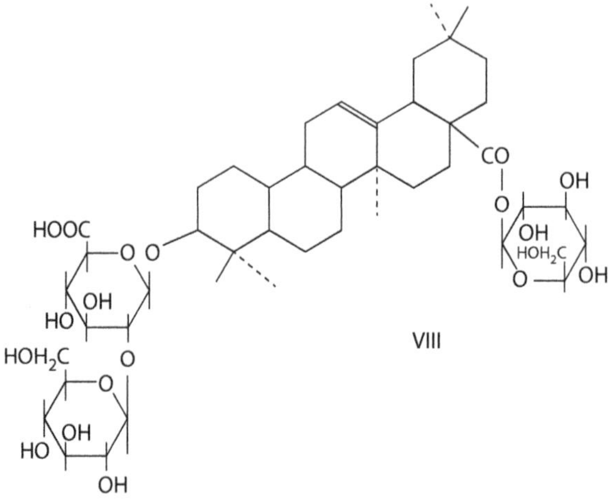

FIGURE 13.4   Ginsenoside Ro (Chikusetsu Saponin V).

Organic Acids

Many organic acids are present in the alcohol extracts of ginseng roots. The most common organic acids are citric, fumaric, ketoglutaric, oleic, linoleic, linolenic, maleic, malic, pyruvic, succinic, tartaric, and other yet unidentified acids.[53,54] As to ginseng's activity, what role these acids play is not known.

Nonprotein Nitrogenous Substances

Nitrogenous substances, about 2%–5% of unknown structure, were detected in the ethereal extract of ginseng root; the nonprotein nitrogenous substances were about 1%–1.5%.[55,56] Although in an early study, it was shown that American ginseng does not contain an alkaloid,[57] a nitrogenous substance called *choline* has been isolated and identified in the alcoholic extract of ginseng. Choline is an alkaloid in nature; it is found in plants and in animal organs. Choline chloride has been used as a lipotropic agent (preventing the excess of accumulation of fat). Choline also gives a marked hypotonic action on blood pressure in rabbits.[58]

Amino Acids and Peptides

Korean white ginseng root contains at least 15 different free amino acids. Some of them are essential, and others are nonessential.[59] Essential amino acids are so called because they must be provided by the diet, since the body does not manufacture them.[60] The essential amino acids found in ginseng are arginine, histidine, lysine, leucine, threonine, valine, and phenylalanine. Nonessential amino acids, alanine, aspartic acid, glutamic acid, glycine, proline, tyrosine, and serine, are also found in ginseng. Some of these amino acids are found in the stems and leaves, as well as in the root.

VITAMINS

The ginseng root contains many biologically essential vitamins. So far, vitamin B complex, biotin, niacin, niacinamide, and pantothenic acid have been identified.[61] The presence of vitamin B12,[62] nicotinic acid, and folic acid has also been reported.[62,63] The contents of these vitamins were shown to vary with the age and origin of the ginseng plant.

MINERALS AND TRACE ELEMENTS

The ginseng root contains many biomedically important mineral and rare element substances. In the water extract of ginseng, elements such as aluminum, manganese, potassium, phosphorus, silicate ions,[64] sulfur, magnesium, calcium, and iron were found.[19] The additional minerals and trace elements found were sodium, zinc, copper, molybdenum, and boron.[65]

Trace amounts of the rare elements of manganese (Mn), vanadium (V), copper (Cu), cobalt (Co), and arsenic (As) contained in Korean ginseng root were detected by a radioactivation method.[66] These minute amounts of rare elements were found to be extremely important to our health. Ginseng roots cultivated in different areas of Korea contain different amounts of these rare elements (Table 13.4).[67]

ENZYMES

Korean fresh ginseng root and dried white root contain diastase, which changes starches into sugars, but the red ginseng root does not.[68] Korean ginseng root also contains other unidentified enzymes and ferments.

GINSENG SAPONINS IN LEAVES AND STEMS

The total crude ginseng saponins vary significantly according to *Panax* species, sources of cultivation, and the parts of the plant extracted. Ginseng leaves, stems, and buds contain about 5%–15% of ginseng saponins in comparison to 2%–15% saponins found in roots. Saponin mixtures found in ginseng roots are called *ginsenoside R*

TABLE 13.4
Korean Ginseng Root Cultivated in Different Areas

Trace Element	Buyo	Kumsan	Kanghwa	Poongki
Manganese	25.9 ppm[a]	19.0 ppm		
Vanadium	0.023 ppm	0.02 ppm		
Copper		7.0 ppm		
Cobalt		0.06 ppm	9.0 ppm	
Arsenic		0.44 ppm	0.06 ppm	0.25 ppm

[a] ppm, parts per million.

(where *R* denotes root), while the saponics found in leaves are called *ginsenoside F* (where *F* represents folia, meaning leaves).[69,70]

Komatsu and Tomimori isolated fatty acids, nonacosane, and phytosterols from the nonvolatile fraction of the ether extract of ginseng leaves, stems, and flowers. Later, a new natural flavonoid (yellow plant pigment) called *panasenoside* was isolated and characterized.[71,72]

Ginseng is well known for its beneficial biological effects on the human body. While the plant contains various ingredients, ginsenosides play a more significant role in exerting pharmacological actions than any other constituents. Of the great number of ginsenosides present in *P. ginseng*, fewer than 10 account for most ginsenoside contents. In particular, ginsenosides Rb_1, Rb_2, Rc, Rd, Re, Rf, and Rg_1 are most abundant in the roots of raw ginseng. Intriguingly, chemical reactions during the processing of ginseng, such as oxidation, hydrolysis, and/or dehydration, lead to the formation of artifactual compounds, which often have enhanced biological activities. Besides, orally administered ginsenosides undergo biotransformations in the gastrointestinal tract, and some metabolites produced by the action of bacteria have structures different from those of naturally occurring ginsenosides.

CHEMICAL CONSTITUENTS OF AMERICAN GINSENG

Pioneer studies in Japan showed that American ginseng roots contain many ingredients similar to those found in Korean and Chinese ginseng, but the American root contains slightly more saponin but less proteinous and oily substances.[76–79]

American ginseng also contains inorganic salts (sulfate and phosphate salts of iron, aluminum, calcium, silicon oxide, and manganese), reducing and nonreducing sugars, phytosterol esters, terpene, panacene, and fatty acids.

Here in the United States, Wong reported his studies on the American ginseng root in 1921. The authentic samples of four and six-year-old American ginseng roots were crushed and extracted with different solvents. A small amount of ginseng oil, sugar, and saponins were obtained but none contained any alkaloid. By using different organic solvents, quite different amounts of total extractives were obtained. A small amount of enzyme was also detected in the extract, but the nature of the enzyme was not identified. The total extracted ginseng oil was about 0.8%.[57] The American ginseng oil was also explored by Torney and Cheng.[73]

In recent years, E. John Staba of the University of Minnesota made extensive studies on American ginseng. He concentrated on the saponins and sapogenins contained in the American ginseng plant. For the first time the saponins of the above-ground parts (stems, leaves, and fruits) of American ginseng were also explored. Jung-yun Kim under the supervision of Staba, earned his Ph.D. from the studies of the phytochemistry of American ginseng in 1974.[74,75] They found that the aboveground parts of the plant contain many of the saponins normally present in the root. The ethereal extract portion contains both β-sitosterol and stigmasterol. The methanol extract contains predominantly ginseng saponins and saccharides. A total of 11 ginseng saponins, called *panaquilins*, were isolated and identified. The term *panaquilin* is similar to *panaquilon*, a name introduced originally by Garriques for the ginseng saponin he isolated from American ginseng root in 1854.[2] These newly isolated American ginseng saponins are panaquilins A, B, C, D, E_1, E_2, E_3, G_1, G_2 (c), and (d). In their studies it was found that the amounts of saponins contained in the ginseng plant vary not only with the part of the plant but also with the age of the plant and the season of collection. For example, a 6-year-old root contains a different saponin mixture from that found in a 2-year-old root. Similarly, the ginseng leaves collected in July contain different saponins from those collected in September. Panaquilins B, C, E_2, E_3, and G_2 are present throughout the plant. Panaquilins D, E_1, and G_1 are present mainly in the root rather than in the above-ground parts; while panaquilins (c), (d), and G_2 are found predominantly in the above-ground parts rather than in the root. The American ginseng leaves contain panaquilins B, C, (d), E_2, E_3, and G_2. After acid hydrolysis of these saponins, sapogenins (aglycones) were obtained.

SECRET PRINCIPLES OF GINSENG

Staba carefully compared the properties of panaquilins with those of ginsenosides (isolated by Shibata et al.) and panaxosides (isolated by Elyakov et al.) and found that the American ginseng root has a different composition of ginseng saponins. Table 13.5 gives a comparison of these ginseng saponins.

Although *P. ginseng* and American ginseng contain similar ginseng saponins, panaquilins E_1 and G_2 (in American ginseng) are absent in the root of *P. Ginseng*, while ginsenosides Ra and Rf are not present in the root of American ginseng. The American ginseng root contains about 17.3% panaxadiol, 0.44% panaxatriol, and

TABLE 13.5

Similarities of Panaquilins, Ginsenosides, and Panaxosides Isolated from American and Korean Ginseng Roots

Plant Species	American Ginseng (*Panax quinquefolium*)	*Panax ginseng*
Panaquilin	Ginsenoside (Japanese Terminology)	Panaxoside (Russian Terminology)
—	Ro	—
—	Ra	F
B	Rb_1, Rb_2	E
C	Rc	D
D	Rd	—
E_1	—	—
E_2	Rd	C
E_3	Re	B
—	Rf	—
G_1	Rg_1	A
G_2	Rg_2	—
—	Rg_3	—
—	Rh_1	—
—	Rh_2	—

0.28% oleanolic acid. The approximate ratio of panaxadiol to panaxatriol of American ginseng root is about 39:1 by column chromatography, or 40:1 by preparative chromatography. The ratio of sapogenin panaxadiol to panaxatriol for *P. ginseng* is about equal. Accordingly, the American ginseng contains much more panaxadiol than *P. ginseng* root.[74,75,79,80] This could be the reason why the Chinese have always said that American ginseng is different in medicinal properties from *Panax ginseng*, or Chinese ginseng.

STRUCTURAL DIVERSITY OF GINSENOSIDES

Ginsenosides belong to a family of steroids with a four trans-ring rigid steroid skeleton. Most ginsenosides share a unique triterpenoid saponin structure of the dammarane type. More than 100 ginsenosides have been isolated from roots, leaves, stems, flower buds, and berries of Asian ginseng and American ginseng, and these ginsenosides exhibit considerable structural variation. Ginsenosides differ from one another by the type of sugar moieties, sugar number, and site of sugar attachment at positions C-3, C-6, or C-20. The structural isomerism and stereoisomerism, the number and site of attachment of hydroxyl groups, and available modified side chain at C-20 also increase their diversity.

Ginsenosides from ginseng are divided into several groups. Protopanaxadiol (PPD) and protopanaxatriol (PPT) groups are the main constituents, while ocotillol and oleanane groups are minor ones. The PPD group has sugar moieties attached to the β-OH at C-3 and/or c-20, and the PPT group has sugar moieties attached to the α-OH at C-6 and/or β-OH at C-20.[4,5,12] The ocotillol group has a five-membered epoxy ring at C-20, and the oleanane group has a modified C-20 side chain.[13]

Chemically, several differences exist between Asian ginseng and American ginseng. An important parameter used for this differentiation is the presence of ginsenoside Rf in Asian ginseng but pseudoginsenoside F11 in American ginseng. High-performance liquid chromatography–evaporative light scattering detector (HPLC-ELSD) or high-performance liquid chromatography–mass spectrometry (HPLC-MS) can be used to detect both F11 and Rf. In addition, ratios of Rg_1/Rb_1 and Rb_2/Rb_1 are useful. Both ratios less than 0.4 is indicative of American ginseng, while a high value of ratios is characteristic of Asian ginseng. One exception is wild American ginseng, which may have a high Rg_1-to-Rb_1 ratio.

Like Asian ginseng, a recent study on American ginseng shows that ginsenoside content also varies among different parts of the plant. The leaf contains the highest ginsenosides (16.5%), followed by root-hair (6.9%), rhizome (5.1%, root (4.9%), and stem (2.0%). The content of ginsenosides increases with the age of the plant parts, except the leaf. In general, ginsenoside Rb_1, Re, Rd, Rc, Rg_1, and Rb_3 are the six major saponins in American ginseng, accounting for more than 70%. Variability in individual ginsenosides and total ginsenoside amount has been observed in different commercial products of American ginseng, which is in part associated with natural variations such as climate, geographical location, and cultivation length and conditions. This ginsenoside variability in different ginseng products may also be responsible for different or even opposing reported pharmacological activities.

Thus, the importance of standardization of ginseng products should be primarily emphasized.

To mimic Korean (or Asian) red ginseng, American red ginseng can be prepared experimentally using steaming or heating treatment (e.g., at 120°C for 4 hours). The chemical composition of the steamed American ginseng is quite different from the untreated ginseng. The steaming process causes obvious chemical degradation and conversion of original saponins to some newly occurring compounds. The polar ginsenosides including Rg_1, Re, Rb_1, Rc, Rb_2, Rb_3, and Rd decrease remarkably, while less polar ginsenosides, including Rg_2, Rg_3, Rg_5, Rh_2, Rk_1, and Rs_4, increase. Due to the change in ginsenoside profile, the steaming treatment may enhance American ginseng's effects, such as increasing its anticancer activities.

CONSTITUENTS OF JAPANESE GINSENG

Japanese ginseng (*Panax japonicum* C. A. Meyer) is also called *chikusetzu ninjim* in Japanese. It has been the most widely used ginseng substitute in Japan, chiefly for stomachic, expectorant, and antipyretic action.

Inuoune[4] seemed to be the first who reported finding and testing pharmacologically a saponin isolated from Japanese ginseng in 1902. Later Murayama et al.,[80,81] Aoyama,[82] and Kotake et al.[83] also investigated the crude ginseng saponin. Kitasota et al.[84] and Kuwata and Matsukawa[85] isolated oleanolic acid from Japanese ginseng. In 1962, Shibata et al. reported that in addition to oleanolic acid, arabinose, glucose, and glucuronic acid, a small amount of panaxadiol was also present in saponin mixture isolated from Japanese ginseng.[28]

Kondo et al. isolated three saponins, *chikusetsu* saponins III, IV, and V, in crystalline forms and also confirmed the structures of these saponins.[86] *Chikusetsu* saponin III is composed of protopanaxadiol or 20-epi-protopanaxadiol, glucoses (2 moles), and xylose. This particular saponin is a homologous saponin of ginsenoside found in *P. ginseng*. *Chikusetsu* saponin IV is composed of oleanolic acid bound with glucose, arabinose, and glucuronic acid. *Chikusetsu* saponin V is also an oleanolic acid bound with glucoses (2 moles) and glucuronic acid. This particular saponin is found to be identical to ginsenoside Ro of ginseng saponin.

The structure of *chikusetsu* saponin IV is shown to be identical to a tonic saponin of araloside A, which was isolated from *Aralia manschuria*, Araliaceae, by Kochetkov et al.[87]

CONSTITUENTS OF SIBERIAN GINSENG

Siberian ginseng (*Eleutherococcus senticosus* of Araliaceae) contains many biologically active saponin glycosides. At least six glycosides, called *eleutherosides*, have been isolated from the root of Siberian ginseng. Eleutherosides A, B, C, D, E, and F are known and have been tested in animals.[88,89]

Eleutheroside A has been identified as *daucosterin*. Eleutherosides B, D, and E are structurally close, and they all are syringarosinol. Eleutheroside C is a nonglycoside, while eleutherosides A, B, C, and E contain aglycones of varied structures. Eleutheroside F is of unknown structure.[90]

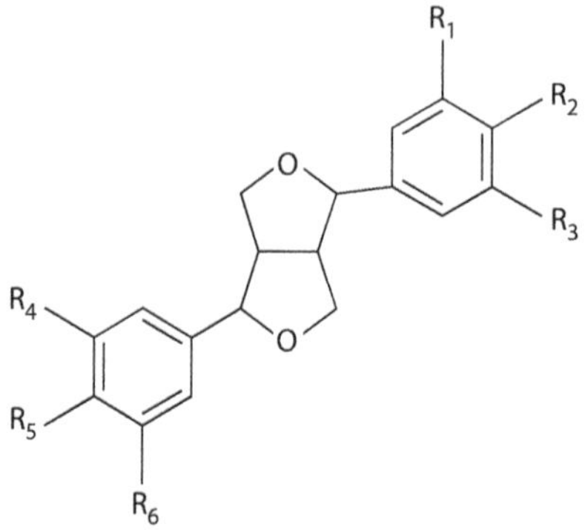

FIGURE 13.5 Phytochemicals and constituents of Siberian ginseng.

The active constituents in eleuthero are known as eleutherosides. The eleutherosides are actually a diverse group of quite different compounds and include the following: lignans—eleutherosides D and E (syringarosides) and B4 (sesame); phenylpropane derivatives including eleutheroside B (syringin), caffeic acid, and chlorogenic acid; coumarins including eleutheroside B1; and isofraxidia sterins including eleutheroside A (daucosterin) (Figure 13.5).

PHYTOCHEMICALS AND CONSTITUENTS OF SIBERIAN GINSENG

Eleuthero contains phenolics, polysaccharides, and eleutherosides A–G, the total content ranging between 0.6% and 0.9% (Table 13.6). Eleutherosides B, B1, and E are representative of three classes of compounds collectively called eleutherosides. Other constituents include phenylpropanoids, lignans, coumarins, polysaccharides, and sugars. A review of the chemistry of eleuthero with 29 chemical structures has been published.

1. Seven primary eleutherosides have been identified, with most of the research attention focusing on eleutherosides B and E.
2. Eleuthero also contains complex polysaccharides (a kind of sugar molecule).
3. These constituents play a critical role in eleuthero ability to support immune function: glycosides (eleutherosides), resins, anthocyanin, and pectin.

The herbal classes CHMM (*Chinese Herbal Materia Medica*) recorded that the root of the Siberian ginseng contains various kinds of natural constituents, including

TABLE 13.6

Phytochemicals and Constituents of Siberian Ginseng

R_1	R_2	R_3	R_4	R_5	R_6
OCH_3	O-β-D-Gle	CH_3	OCH_3	O-β-D-Gle	OCH_3

"eleutheroside A, eleutheroside B, eleutheroside C, eleutheroside D, eleutheroside E, eleutheroside A is daucosterol, eleutheroside B is syringin; the root also contains sinapaldehyde glucoside, coniferaldehyde glucoside, coniferin, liriodemdrin, amygdlin, [O-α-L-rhamnopyranosyl-(1→4)-[O-α-L-rhamnopyranosyl-(1→2)]-O-β-D-glucoronopyranosyl]-16-hydroxy-13β,28-epoxyoleanane], etc."

PANAX NOTOGINSENG

Panax notoginseng is a species of the genus *Panax*, and it is most commonly referred to in English as notoginseng. In Chinese it is called *tiánqī* ( ), *tienchi ginseng, sānqī* ( ) or *sanchi*, "three-seven root," and "mountain paint." Notoginseng belongs to the same scientific genus as Asian ginseng. In Latin, the word *panax* means "cure-all," and the family of ginseng plants is one of the best-known herbs.

Notoginseng grows naturally in China and Japan. The herb is a perennial with dark-green leaves branching from a stem with a red cluster of berries in the middle. It is both cultivated and gathered from wild forests, with wild plants being the most valuable. The Chinese refer to it as "three-seven root" because the plant has three petioles with seven leaflets each. It is also said that the root should be harvested between 3 and 7 years after planting it.

It is classified in Chinese medicine as warm in nature, sweet and slightly bitter in taste, and nontoxic. The dose in decoction for clinical use is 5–10 g. It can be ground to powder for swallowing directly or taken mixed with water: the dose in that case is usually 1–3 g. In the *Bencao Gangmu* (*Compendium of Materia Medica*, AD 1596), it is stated: "On account of the fact that sanqi is a herb belonging to the blood phase of the yang ming and jue yin meridians, it can treat all diseases of the blood." Notoginseng is an herb that has been used in China quite extensively since the end of the nineteenth century. It has acquired a very favorable reputation for treatment of blood disorders, including blood stasis, bleeding, and blood deficiency. It is the largest ingredient in Yunnan Bai Yao, a famous hemostatic proprietary herbal remedy that was notably carried by the Viet Cong to deal with wounds during the Vietnam War.

Notoginseng, *P. ginseng, P. quinquefolius*, and *P. vietnamensis*, contain dammarane-type ginsenosides as the major constituents. Dammarane-type ginsenosides include two classifications: the 20(S)-protopanaxadiol (ppd) and 20(S)-protopanaxatriol (ppt) classifications. *P. notoginseng* contains high levels of Rb_1, Rd (ppd classification), and Rg_1 (ppt classification) ginsenosides. Rb_1, Rd, and Rg_1 content of *P. notoginseng* is found to be higher than that of *P. ginseng* and *P. quinquefolius* in one study.

Eight saponins, namely, saponins A, B, C, D, E, F, G, and H, have been successfully isolated, and among them saponin A and saponin D have been identified.[91]

Hydrolysis of saponin A, a white powder, in dilute HC1 gives oleanolic acid, glucose, and glucuronic acid. From chemical and physical data, it has been confirmed that saponin A is identical to *chikusetsu* saponin V, the main saponin of *P. japonicum* C. A. Meyer.

Saponin D, also, is a white powder, and the IR spectra showed the presence of hydroxyl and olefinic bonds. The acid hydrolysis or saponin D with HC1-dioxane-water gave panaxadiol and glucose, which were identified by TLC, paper partition chromatography, and gas liquid chromatography. It consists of 20-protopanaxadiol and 4 moles of glucose. Saponin D has been confirmed to be identical to ginsenoside Rb_1, the chief saponin present in ginseng root.

The hydrolyzed products of saponins showed some differences among the different ginseng species. Japanese and Himalayan ginseng roots contain mainly oleanane-type triterpenes, while ginseng and American ginseng roots contain more dammarane-type triterpenes glycosides.[79]

It is very interesting to note that the leaves of many *Panax* species such as Chinese ginseng, Japanese ginseng, Himalayan ginseng, as well as American ginseng, contain a certain amount of many of the saponins present in the roots.[69,74] The Chinese ginseng leaves contain more panaxadiol and panaxatriol but little oleanolic acid. The leaves of Japanese ginseng contain more panaxatriol and oleanolic acid but little panaxadiol, while Himalayan ginseng leaves contain more panaxadiol.[79]

To investigate the chemical constituents of the processed rhizomes of *Panax notoginseng*, their 70% ethanol extract was chromatographed on microporous resin (SP825), silica gel, rP-C18, and semipreparative HPLC to afford compounds 1–23. On the basis of physicochemical properties and spectral data analysis, their structures were identified to be $6'-O$-Acetylginsenoside Rh_1, ginsenoside Rk_3, ginsenoside Rh_4, 20s-ginsenoside Rg_3, ginsenoside Rk_1, 20R-ginsenoside Rg_3, ginsenoside Rg_5, ginsenoside F_2, 20s-ginsenoside Rh_1, 20R-ginsenoside Rh_1, gypenoside X VII, notoginsenoside Fa, ginsenoside Ra_3, ginsenoside Rg_1 (14), ginsenoside Re (15), notoginsenoside R_2 (16), ginsenoside Rg_2, notoginsenoside R_1, ginsenoside Rd, ginsenoside Rb_1, notoginsenoside D, notoginsenoside R_4, and ginsenoside Rb_2, respectively. Among them, compound 1 was isolated from *P. notoginseng* for the first time, and compounds 4, 6, 8, and 11 were isolated from the processed *P. notoginseng* for the first time. According to the fingerprint profiles of raw and processed *P. notoginseng*, the putative chemical conversion pathways of panoxatriol and panoxadiol compounds in the processing procedure were deduced, and the results revealed the main reactions to be dehydration and glycosyl hydrolysis.

REFERENCES

1. Garriques, S. S., *Am. J. Pharm.*, 26: 511, 1854.
2. Garriques, S. S., *Ann. Chem. Pharm.*, 90: 231, 1854.
3. Davydow, *Pharm. Zeitscher, Russland*, 29: 97, 1889.
4. Inoue, M., *J. Pharm. Soc., Japan*, 242: 326, 1902.
5. Fujitani, K., *J. Tokyo Med. Assoc.*, 2: 43, 1905.
6. Asahina, U. and B. Taguchi, *J. Pharm. Soc., Japan*, 25: 547, 1906.
7. Kondo, H. and G. Tanaka, *J. Pharm. Soc., Japan*, 35: 779, 1915.
8. Kondo, H. and S. Yamaguchi, *J. Pharm. Soc., Japan*, 38: 747, 1918.
9. Kondo, H. and U. Amano, *J. Pharm. Soc., Japan*, 40: 1027, 1920.
10. Abe, K. and I. Saito, *Japan Med. World*, 2: 166, 1922.
11. Yonekawa, M., *Japan Med. World*, 2: 785, 1926.
12. Kotake, M., *J. Chem. Soc., Japan*, 51: 357, 1930; 51: 396, 1930.
13. Sakai, W., *Tokyo Igakukai Zasshi*, 31: 1, 1917.
14. Sakai, W., *Japan Med. Lit.*, 3: 27, 1918.
15. Sakai, W., *Ijishimpun*, 990: 112, 1918.
16. Sakai, W., *Japan Med. Lit.*, 5: 6, 1920.

17. Shibata, S., *Tampakushitsu*, 12: 32, 1967.
18. Tanaka, O., *Taisha*, 10: 548, 1973.
19. Han, B. H., in *"Korean Ginseng Symposium,"* Korean Society of Pharmacognosy, 1974, pp. 81–111.
20. Claus, E. P., V. E. Tayler, and L. R. Brady, *Pharmacognosy*, 6th ed., Lea and Febiger, Philadelphia, Chapter 4, 1970.
21. Trease, G. E. and W. C. Evans, *Pharmacognosy*, Williams and Wilkins, Baltimore, Chapter 12, 1972.
22. Brekhman, I. I., "Panax Ginseng," *Medgiz* (Leningrad): 182, 1957.
23. Naidenova, I. N., V. A. Andreeva, V. T. Bykov, S. P. Versen, E. S. Zyakhov, and V. F. Chernil, *Izvest. Acad. Nauk*, USSR, Otdel Khim Nauk: 1403, 1957.
24. Elyakov, G. B., L. I. Strigina, A. J. Khorlin, and N. K. Kochetkov, *Izvest. Acad. Nauk*, S.S.S.R., Ser. Khim Nauk: 1125: 2054, 1962.
25. Elyakov, G. B., L. I. Strigina, N. I. Uvarova, V. E. Vskovsky, A. K. Dzizenko, and N. K. Koochetkov, *Tetrahedron Lett.*, 48: 3591, 1964.
26. Elyakov, G. B., A. K. Dzizenko, and Yu. N. Elkin, *Tetrahedron Lett.*, 1: 141, 1966.
27. Andreev, L. V., L. I. Slepyan, and I. K. Nikitina, *Rast. Resur.*, 10: 126, 1974.
28. Fujita, M., H. Itokawa, and S. Shibata, *J. Pharm. Soc. Japan*, 82: 1634–1638, 1962.
29. Shibata, S., M. Fujita, H. Itokawa, O. Tanaka, and T. Ishii, *Tetrahedron Lett.*, 10: 419, 1962.
30. Shibata, S., O. Tanaka, M. Nagai, and T. Ishii, *Tetrahedron Lett.*, 26: 1239, 1962.
31. Shibata, S., O. Tanaka, M. Nagai, and T. Ishii, *Chem. Pharm. Bull.*, 11: 762, 1963.
32. Shibata, S., M. Fujita, H. Itokawa, O. Tanaka, and T. Ishii, *Chem. Pharm. Bull.*, 11: 759, 1963.
33. Shibata, S., O. Tanaka, M. Sado, and S. Tsushima, *Tetrahedron Lett.*, 12: 795, 1963.
34. Tanaka, B., M. Nagai, and S. Shibata, *Tetrahedron Lett.*, 33: 2291, 1964.
35. Shibata, S., O. Tanaka, K. Some, Y. Iida, T. Ando, and H. Nakamura, *Tetrahedron Lett.*, 3: 207, 1965.
36. Sanada, S. et al., *Chem. Pharm. Bull.*, 22: 421, 2407, 1974.
37. Shibata, S. et al., *Chem. Pharm. Bull.*, 14: 595, 1966.
38. Iida, Y., O. Tanaka, and S. Shibata, *Tetrahedron Lett.*, 52: 5449, 1968.
39. Wagner-Jauregg, T. and M. Roth, *Pharm. Acta. Helv.*, 38: 125, 1963.
40. Lin, Y. T., *J. Chin. Chem. Soc.*, (Taiwan), 8: 109, 1961.
41. Horhammer, L., H. Wagner, and B. Lay, *Pharm. Ztg.*, 106: 1307, 1961.
42. Takahashi, M., K. Isoi, M. Yoshikura, and T. Osugi, *J. Pharm. Soc., Japan*, 81: 771, 1961.
43. Takahashi, M. et al., *J. Pharm. Soc., Japan*, 84: 752, 757, 1964.
44. Takahashi, M. and M. Yoshikura, *J. Pharm. Soc., Japan*, 86: 1051, 1053, 1966.
45. Euler, H. and E. Nordenson, *Z. Physiol. Chem.*, 56: 223, 1908.
46. Ahn, Y. P. and C. C. Chung, *Kor. J. Chem. Soc.*, 14: 281, 1970.
47. Takiura, K. and I. Nakagawa, *J. Pharm. Soc., Japan*, 83: 298, 1963.
48. Takiura, K. and I. Nakagawa, *J. Pharm. Soc., Japan*, 83: 301, 1963.
49. Takiura, K. and I. Nakagawa, *J. Pharm. Soc., Japan*, 83: 305, 308, 1963.
50. Ovodov, Y. S. and T. F. Solov'eva, *Khim. Prir. Soedin.*, 2: 299, 1966.
51. Solov'eva, T. F., L. V. Arsenyuk, Y. S. Ovodov, *Carbohd. Res.*, 10: 13, 1969.
52. Solov'eva, T. et al., *Khim. Prir. Soedin.*, 5: 201, 1969.
53. Park, M. S., *In-Sam-Mun-Hun-Teuk-Zip*, Central Technical Research Institute, the Office of Monopoly, Seoul, 1969, p. 1.
54. Lee, C. Y. and T. Y. Lee, in *"Sym. Photochem. Proc. Meeting,"* University of Hong Kong, 1961, p. 171.
55. Gstirner, F. and H. J. Vogt, *Arch. Pharm.* (Weinheim), 300: 371, 1967.
56. Gstirner, F. and W. Brown, *Arch. Pharm.* (Weinheim), 296: 384, 1963.

57. Wong, Y. C., *J. Am. Pharm, Assoc.*, 10: 431, 1921.
58. Takatori, K. et al., *Chem. Pharm. Bull.*, 11: 1342, 1963.
59. Gstirner, F. and H. J. Vogt, *Arch. Pharm.* (Weinham), 299: 936, 1966.
60. White, A., P. Handler, and E. I. Smith, *Principles of Biochemistry*, 4th ed., McGraw-Hill, New York, 1968, p. 540.
61. Goto, M., *J. Pharm. Soc., Japan*, 77: 467, 471, 1957.
62. Kim, Y. E., S. K. Juhn, and B. J. An, *Yakhak Hoeji (J. Pharm. Soc., Korea)*, 8: 80, 1964.
63. Kim, Y. E. and M. O. Her, *Yakhak Hoeji (J. Pharm. Soc., Korea)*, 8: 85, 1964.
64. Nomura, S. and Y. Oshima, *J. Chosen Med. Assoc.*, 20: 3, 1930.
65. Jhang, J. J., E. J. Staba, and J. Y. Kim, *In Vitro*, 9: 253, 1974.
66. Pijck, J. and J. I. Kim, *J. Pharm. Belg.*, 19: 3, 1964.
67. Pijck, J. and A. Claeys, "Mededel." Landbouwhogeschool Opzoekingssta, *Staat. Gent.*, 30: 1295, 1965.
68. Yamaguchi, I., *J. Chosen Med. Assoc.*, 125(5), 568, 1928.
69. Yoshizaki, M. et al., *Shoyakugaku Zasshi*, 27: 110, 1973.
70. Namba, T. et al., *J, Pharm. Soc., Japan*, 94: 252, 1974.
71. Komatsu, M. and T. Tomimori, *Shoyakugaku Zasshi*, 20: 21, 1966.
72. Komatsu, M. and Y. Makiguchi, *J. Pharm. Soc., Japan*, 89: 122, 1969.
73. Torney, H. J. and F. M. Y. Cheng, *St. Bonaventure Sci. Studies*, 7: 9, 1939.
74. Kim, J. Y. and E. J. Staba, in *"Proceedings of International Ginseng Symposium,"* Central Research Institute, Office of Monopoly, Republic of Korea, 1974, pp. 77–93.
75. Kim, J.-Y., "Saponin and Sapogenin Distribution in American Ginseng Plant," *Ph.D. thesis*, 1974, University of Minnesota, Xerox University Microfilm, Ann Arbor, 1975.
76. Nomura, S. and Y. Oshima, *J. Chosen Med. Assoc.*, 21: 558, 1931.
77. Nomura, S. and Y. Oshima, *J. Chosen Med. Assoc.*, 21: 553, 1931.
78. Min, P. A., *Folia Pharmacol, Japan*, 11: 238, 256, 1931.
79. Shibata, S. et al., *J. Pharm. Soc., Japan*, 85: 753, 1965.
80. Murayama, Y. and T. Itagaki, *J. Pharm. Soc., Japan*, 43:783, 1923.
81. Murayama, Y. and T. Itagaki, *J. Pharm. Soc., Japan*, 47: 526, 1927.
82. Aoyama, S., *J. Pharm. Soc., Japan*, 50: 1065, 1930.
83. Kotake, M., *J. Pharm. Soc., Japan*, 50: 396, 1930.
84. Kitasota, I. and S. C. Sone, *Acta Phytochim., Japan*, 6: 179, 1932.
85. Kuwata, S. and Matsukawa, *J. Pharm. Soc., Japan*, 54: 211, 1934.
86. Kondo, N. et al., *Chem. Pharm. Bull.*, 18: 1558, 1970; 19: 1103, 1971.
87. Kochetkov, N. K., A. J. Khorlin, and V. E. Vaskovsky, *Tetrahedron Lett.*, 713, 1962.
88. Brekhman, I. I. and I. V. Dardymov, *Lloyda*, 32: 46, 1969.
89. Brekhman, I. I., *Eleutherococcus*, Nauka Publishing House, Leningrad, 1968.
90. Ovodov, Y. S. et al., *Khim. Prirodnih Soedin*, 1: 63, 1967.
91. Kondo, N., J. Shoji, and O. Tanaka, *Chem. Pharm. Bull.*, 21: 2702, 1973.

14 How Differences in Chemical Constituents in Ginseng Affect Medicinal Effects

It has been confirmed that all pharmacologic and medicinal active components in ginseng plants are mostly from saponins. As of 1999, it was reported that 34 ginsenosides have been identified in *Panax ginseng* roots, the dominant saponins in these roots are of the Rb series and are of the dammarane type. Most of the other saponins are present in miniscule amounts (the next largest group, the Rg series, also of the dammarane group, are present at about one-third the quantity of the Rb series in *Panax ginseng*.

The ginsenosides are concentrated primarily in the root's cortical tissues (outer layers) compared to the quantities found in the interior portions. According to one Chinese report, the ginsenoside content of the root hairs (fibrous lateral roots that are mainly cortical-type material) was 9.7%, while that in the thicker lateral roots was 6.4%, and that in the main tap root (which is the portion usually traded on the market, having little cortical material) was only 3.3%. An earlier European report gave the figures as 5%–7%, 2%–4%, and 0.7%–1.7% for these ginseng root parts, respectively.

The root of *P. ginseng* contains a large number of different constituents. There are primarily a number of *saponins*, generally referred to as ginsenosides. The three genuine sapogenins of ginsenosides were identified as 20(S)-protopanaxadiol, 20(S)-protopanaxatriol, and oleanolic acid.[1–3] Panaxadiol and panaxatriol, which were supposed to be the sapogenins, were later found as artifacts during acid hydrolysis of saponins.[3–5] The sapogenins and saponins isolated from ginseng root are listed in Table 14.1.[6–14]

The only oleanolic acid-type saponin so far identified in the root is ginsenoside Ro, or 3β-*O*-D-glucopyranosyl-(1→2)-*O*-β-D-glucoyranosiduronyl-oleanolic acid 28-β-D-glucopyranosyl ester.[8] Thus, ginsenosides Rc, Rg_1 and Ro are the representatives of the ginseng saponins of protopanaxadiol, protoanaxatriol and oleanolic acid type, respectively. 20(*R*)-Ginsenosides Rg_2 and Rh_1, ginsenosides Rg_3, Rh_2, Rs_1, and Rs_2 are only found in red ginseng. They are believed to be the degradation products of the processing of red ginseng.[11,15]

Other principles isolated from the root include acetylenic compounds panaxynol, panaxydol, panaxytriol, and heptadeca-1-ene-4,6-diyn-3,9-diol.[16–20] From the volatile fraction of the root, a series of sesquiterpenes such as eremophilene, β-gurjunene, *trans*- and *cis*-caryophyllene, ε-muurolene, γ-patchoulene, β-eudesmol, β-farnesene, β-bisabolene, aromadendrene, alloaromadendrene, β-guaiene, γ-elemene, mayurone, pentadecane, and 2,5-dimethyltridecane were also isolated.[21] Some pyrazine

TABLE 14.1

Saponins from *Panax Ginseng* Root

Classification	Saponins
Protopanaxadiol-type ginsenosides	Ginsenoside Ra_1
	Ginsenoside Ra_2
	Ginsenoside Ra_3
	Ginsenoside Rb_1
	Ginsenoside Rb_2
	Ginsenoside Rb_3
	Ginsenoside Rc
	Ginsenoside Rd
	Ginsenoside Rg_3
	Notoginsenoside R_4
	Notoginsenoside Rh_2
	Notoginsenoside Rs_1
	Notoginsenoside Rs_2
	Quinquenoside R_1
	Malonylginsenoside Rb_1
	Malonylginsenoside Rb_2
	Malonylginsenoside Rc
	Malonylginsenoside Rd
Protopanaxatriol-type ginsenosides	Ginsenoside Re
	Ginsenoside Rf
	20-Glucoginsenoside Rf
	20-Glucoginsenoside Rg_1
	20(S),20(R)-Glucoginsenoside Rg_2
	20(S),20(R)-Glucoginsenoside Rh_1
	Notoginsenoside R_1

derivatives including 3-*sec*-butyl-2-methoxy-5-methylpyrazine and 3-isopropyl-2-methoxy-5-methylpyrazine were identified in the basic fraction of the volatile extract of the root (Table 14.1).[22]

Furthermore, a number of hypoglycemic peptide glycans named panaxans A-L and panaxans Q-U, were also isolated from the root.[23–26]

The ginsenosides are considered to be the major pharmacologically active agents in ginseng. These are dammarane saponins and can be divided into two classes: the protopanaxatriol class consisting mainly of Rg_1, Rg_2, Rf, and Re and the protopanaxadiol class consisting mainly of Rc, Rd, Rb_1, and Rb_2. Ginseng root also contains other saponins, glycans, and essential oils. Many other constituents have been identified (Tables 14.1 and 14.2).[27]

Ginseng root has been used for thousands of years in the traditional medical system in oriental countries. It occupies a prominent position on the list of the best-selling medicinal plants in the world. Asian ginseng (*Panax ginseng* C. A. Meyer) and American ginseng (*Panax quinquefolius* L.) are the two most recognized ginseng

TABLE 14.2
Distribution of Ginsenosides in *Panax ginseng*

	Percent (%) Content								
	Rg_1	Re	Rf	Rg_2	Rb_1	Rc	Rb_2	Rd	Total
Leaves	1.078	1.524	—	—	0.184	0.736	0.553	1.113	5,188
Leafstalks	0.327	0.141	—	—	—	0.190	—	0.107	0.765
Stem	0.292	0.070	—	—	—	—	0.397	—	0.759
Main root	0.379	0.153	0.092	0.023	0.342	0.190	0.131	0.038	1.348
Lateral roots	0.406	0.668	0.203	0.090	0.850	0.738	0.434	0.143	3.532
Root hairs	0.376	1.512	0.150	0.249	1.351	1.349	0.780	0.381	6.148

botanicals around the world. Compared to the long history of use and the copious amounts of research on Asian ginseng, the study of American ginseng and its constituents is much less extensive.

Table 14.1 and Figure 14.1 show the chemical differences in American ginseng and Asian ginseng. As one of the best-selling herbs in the United States, American ginseng is grown in the eastern temperate forest areas of North America, from southern Quebec, Minnesota, and Wisconsin in the north, to Oklahoma, the Ozark Plateau, and Georgia in the south. With the widespread popularity of herbal medicines in the West, the past few decades have witnessed some promising advances in research on American ginseng and its constituents. The triterpenoid saponins, called ginseng saponins or ginsenosides, are the major active constituents

Protopanaxatriol class   Protopanaxadiol class

	R	R^1
(20)S-protopanaxatriol	H	H
Ginsenoside Re	Glc(2→1)Rha	Glc
Ginsenoside Rf	Glc(2→1)Glc	H
Ginsenoside Rg_1	Glc	Glc
Ginsenoside Rg_2	Glc(2→1)Rha	H

	R	R^1
(20)S-protopanaxadiol	H	H
Ginsenoside Rb_1	Glc(2→1)Glc	Glc(6→1)Glc
Ginsenoside Rb_1	Glc(2→1)Glc	Glc(6→1)Ara p
Ginsenoside Rc	Glc(2→1)Glc	Glc(6→1)Ara f
Ginsenoside Rd	Glc(2→1)Glc	Glc

Ara = α-L-arabinose, Glc = β-D-glucose, Rha = α-L-rhamnose, f = furanose, p = pyranose

FIGURE 14.1 Protopanaxatriol and protopanaxadiol classes.

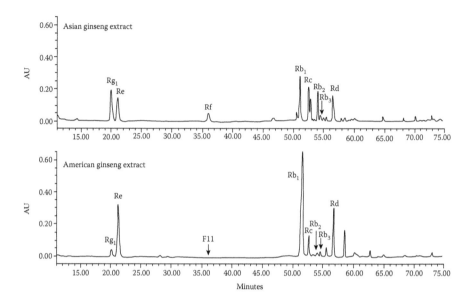

FIGURE 14.2   Asian and American ginseng.

in American ginseng. As shown in Figure 14.1 and Table 14.1, however, American ginseng has a ginsenoside profile different from that of Asian ginseng in terms of total ginsenosides, the ratio of protopanaxadiol (PPD) to protopanaxatriol (PPT), and other marker ginsenosides. In addition, although low Rg_1/high Re was reported in most populations of American ginseng, the high Rg_1/low Re chemotype was also observed (Figure 14.2).

GINSENOSIDE CONTENT COMPARISON

Saponins—plant chemicals—in ginseng are called ginsenosides. While American ginseng contains more ginsenosides than Asian ginseng—29 versus 20 individual ginsenosides—Asian ginseng is more effective medicinally, according to the British Columbia Ministry of Agriculture and Food. This is because the type of ginsenoside best suited for medicinal purposes is available in greater quantities in Asian ginseng. American ginseng contains more of the ginsenosides better suited to adapt to and cope with stress. For both types of ginseng, the age, growing conditions, genetic makeup of the seeds, and where and how it was grown affect the number of ginsenosides available.

Higher levels of ginsenosides than found in *Panax ginseng* (with a differing mix of individual subtypes) are reported for American ginseng (*Panax quinquefolius*), at 6.2%–7.4%, and for Tienchi ginseng (san qi) of southern China (*Panax pseudoginseng*), at 3%–8% ginsenosides, with some reported levels of 12%. In a study of ginsenosides in ginseng roots found on the Taiwanese market, it was reported that the highest content of ginsenosides was found in *Panax pseudoginseng*, followed, in order by

P. quinquefolius, *P. ginseng* root hair, and then red and white *P. ginseng* roots (tap roots, as commonly sold in stores).

Studies of ginseng processing, in which the roots are steamed soon after collection, indicate that red ginseng (often referred to in the literature as Radix Ginseng Rubra), usually has a higher ginsenoside content than white ginseng. During the process to make red ginseng, malonyl ginsenosides are converted to their corresponding ginsenosides by hydrolysis. According to a report on American ginseng, the steaming process used to produce red American ginseng increases the level of its ginsenoside Rb_1 by as much as 1,005 through conversion of malonyl-Rb_1 to Rb_1. The acidic malonyl compounds are poorly absorbed in humans, but intestinal bacteria metabolize malonyl ginsenosides to neutral ginsenosides, which are better absorbed. A small amount of acetyl ginsenosides are also generated from the malonyl ginsenosides when preparing red ginseng. It has been proposed that some of the changes in ginsenosides that occur when white ginseng is steamed to produce red ginseng also arise during preparation of ginseng tea and ginseng extracts. Another factor influencing the difference in ginsenoside content between white and red ginseng is the removal or retention of the outer root skin. White ginseng is frequently prepared by peeling the root. Since the ginsenoside content is particularly rich in the peel, this processing results in a relatively low ginsenoside content. Red ginseng, on the other hand, is processed by steaming the unpeeled roots.

MEDICINAL EFFECTS OF GINSENG SAPONINS

According to traditional Chinese medicine (TCM), Asian ginseng tastes sweet and a little bitter; is warm in property and nature; acts on the lungs, spleen, and stomach meridians; and nourishes the Qi and Shen. The pharmacologic effects and medicinal uses of Asian ginseng are as follows:

- It has been associated with improving overall mental health and ability, particularly in the area of memory.
- It is usually used to invigorate the body's essence, for tiredness, poor appetite, palpitations, mental fatigue, and spontaneous perspiration.
- For qi deficiency of the middle jaw.
- Strengthens the function of the lung-qi.
- Improves type-2 diabetes and polyuria thirst order.
- Strengthens the heart-qi and spleen-qi and calms the spirit. For instability, insomnia, dreaminess, palpitation, listlessness, and lassitude.
- Invigorate kidney-yang. For male sexual disorders, impotence, premature ejaculation, and spermatorrhea, particularly the use of red ginseng.
- Enhances mental capacity.
- Has immunomodulating activity.
- A Mayo Clinic study reviewed that ginseng showed good results in helping cancer patients with fatigue.
- Ginseng has been popular to prevent senility. Its strengthening and energizing effects are important to keep a weak body active well into old age.

American ginseng tastes sweet and a little bitter; is cold in nature; and acts on the heart, lung, and kidney meridians. It nourishes the yin of the body. Pharmacologic effects and medicinal uses are as follows:

- It is associated with lowering blood sugar levels.
- It is used to improve mental health and function as well as act as a mood enhancer and booster.
- It is used to treat attention deficit hyperactivity.
- It nourishes the stomach and yin and benefits qi to treat a dry mouth and tongue.
- It promotes the production of body fluids and reduces internal heat for febrile diseases, lassitude, thirst, night sweating, and fatigue.
- It has the possibility to reduce the risk of cancer.
- It is used to treat colds and flu.
- It has stronger antioxidant activity than Asian ginseng.

In conclusion, according to TCM, the *yin* and *yang* concept views the world as a balance between two opposite and complementary forces. One is rather active and issuing, *yang*; the other is more passive and receptive, *yin*. The Chinese TCM also considers *yin* to be a cooling force and *yang* to be a hot one. The theory that strongly influences traditional medicine is that in the world (as in the human body), a balance should be maintained between *yin* and *yang* forces. Illness is the result of an excess of one of them. In TCM American ginseng supports the *yin* forces, whereas Asian ginseng promotes *yang*. Their activities thus differ and are complementary.

It is nice to see ginseng has been for centuries a trusted herbal remedy, getting the recognition it deserves. Ginseng is called an *adaptogen*, which is a substance that is supposed to help the body better cope with mental and physical stress.

Ginseng has a long history of use in Asia, which has generated interest in continuing the tradition into the present and studying, by modern methods, the constituents of ginseng and their pharmacologic and clinical effects. A careful reading of both the traditional literature and modern Asian and Western research efforts is essential to help guide consumers and health-care professionals toward knowing the correct indications and dosages for ginseng, selecting appropriate ginseng materials to use, and avoiding unsubstantiated claims of beneficial or harmful effects.

REFERENCES

1. Shibata, S., O. Tanaka, T. Ando, M. Sado, S. Tsushima, and T. Ohsawa, "Chemical studies on the oriental plant drugs. XIV. Protopanaxadiol, a genuine sapogenin of ginseng saponins," *Chem. Pharm. Bull.*, 14: 595–600, 1966.
2. Tanaka, O., M. Nagai, and S. Shibata, "The stereochemistry of protopanaxadiol, a genuine sapogenin of ginseng," *Tetrahedron Lett.*, 5: 2291–2297, 1964.
3. Tang, W. and G. Eisenbrand, *Chinese Drugs of Plant Origin—Chemistry, Pharmacology, and Use in Traditional and Modern Medicine*, Springer-Verlag, Berlin, Heidelberg, 1992, pp. 711–737.

4. Shibata, S., M. Fujita, H. Itokawa, O. Tanaka, and T. Ishii, "Studies on the constituents of Japanese and Chinese crude drugs. XI. Panaxadiol, a sapogenin of ginseng roots (1)," *Chem. Pharm. Bull.*, 11: 759–761, 1963.

5. Shibata, S., O. Tanaka, K. Soma, Y. Iida, T. Ando, and H. Nakamura, "Studies on saponins and sapogenins of ginseng. The structure of panaxatriol," *Tetrahedron Lett.*, 207–213, 1965.

6. Besso, H., R. Kaisai, Y. Saruwatari, T. Fuwa, and O. Tanaka, "Ginsenoside-Ra$_1$ and ginsenoside Ra$_2$, new dammarane-saponins of ginseng root," *Chem. Pharm. Bull.*, 30: 2380–2385, 1982.

7. Matsuura, H., R. Kasai, O. Tanaka, Y. Saruwatari, K. Kunihiro, T. Fuwa, "Further studies on dammarane-saponins of ginseng root," *Chem. Pharm. Bull.*, 32: 1188–1192, 1984.

8. Sanaka, S., N. Kondo, J. Shoji, O. Tanaka, and S. Shibata, "Studies on the saponins of ginseng. I. Structures of ginsenoside-Ro, -Rb$_1$, -Rb$_2$, -Rc and -Rd," *Chem. Pharm. Bull.*, 22: 421–428, 1974.

9. Kasai, R., M. Besso, O. Tanaka, Y. Saruwatari, and T. Fuwa, "Saponins of red ginseng," *Chem. Pharm. Bull.*, 31: 2120–2125, 1983.

10. Sanada, S. and J. Shoji, "Studies on the saponins of ginseng. III. Structures of ginsenoside-Rb$_3$ and 20-glucoginsenoside-Rf," *Chem. Pharm. Bull.*, 26: 1694–1697, 1978.

11. Kitagawa, I., M. Yoshikawa, M. Yashihara, T. Hayashi, and T. Taniyama, "Chemical studies on crude drug processing. I. On the constituents of Ginseng Radix Rubra," *Yakygaku Zasshi*, 103: 612–622, 1983.

12. Kitagawa, I., T. Taniyama, T. Hayashi, and M. Yoshikawa, "Malonyl-ginsenoside-Rb$_1$, Rb$_2$, Rc and Rd, four new malonylated dammarane-type triterpene oligosaccharides from Ginseng radix," *Chem. Pharm. Bull.*, 31: 3353–3356, 1983.

13. Sandra, S., N. Kondo, J. Shoji, O. Tanaka, and S. Shibata, "Studies on the saponins of ginseng. II. Structures of ginsenoside-Re, Rf, and Rg$_2$," *Chem. Pharm. Bull.*, 22: 2407–2412, 1974.

14. Ijda, Y., O. Tanaka, S. Shibata, "Studies on saponins of ginseng: The structure of ginsenoside-Rg$_1$," *Tetrahedron Lett.*, 9: 5449–5453, 1968.

15. Li, X. G., "Changes of chemical constituents in ginseng root after processing," *Bull. Chinese Mater. Med.*, 11: 194–199, 1986.

16. Takahashi, M. and M. Yoshikura, "Studies on the components of *Panax ginseng* C. A. Meyer. IV. on the structure of a new acetylene derivative "Panaxynol" (2). Synthesis of 1,7,9-heptadecatrien-4-yn-3-nol," *Yakugaku Zasshi*, 86: 1051–1053, 1966.

17. Takahashi, M. and M. Yoshikura, "Studies on the components of *Panax ginseng* C. A. Meyer. V. on the structure of a new acetylene derivative, 'Panaxynol' (2). Synthesis of 1,9-(*cis*)-heptadecatrien-4,6-diyn-3-nol," *Yakugaku Zasshi*, 86: 1053–1056, 1966.

18. Poplawski, J., J. T. Wrobel, and T. Glinka, "Panaxydol, a new polyacetylenic expoxide from *Panax ginseng* root," *Phytochemistry*, 19: 1539–1541, 1980.

19. Shim, S. C., H. Y. Koh, and B. H. Han, "Polyacetylenes from *Panax ginseng* roots," *Phytochemistry*, 22: 1817–1818, 1983.

20. Dabrowski, Z., J. T. Wrobel, and K. Wojtasiexiez, "Structure of an acetylenic compound from *Panax ginseng*," *Phytochemistry*, 19: 2464–2465, 1980.

21. Zang, H. X., Y. X. Sun, S.Q. Wang, W. P. Jiang, and L. R. Yang, "Analysis of the volatile constituents of Jilin Ginseng," *Kexue Tongbao*, 30: 195–199, 1985.

22. Iwabuchi, H., M. Yoshikura, S. Obata, and W. Kamisako, "Studies on the aroma constituents of crude drugs. 1. On the aroma constituents of Ginseng Radix," *Yagugaku Zasshi*, 104: 951–958, 1984.

23. Konno, C., K. Sugiyama, M. Kano, M. Takahashi, and H. Hikimo, "Validity of the Oriental medicines. LXX. Antidiabetes drugs. 1. Isolation and hypoglycemic activity of panaxans A, B, C, D and E, glycans of *Panax ginseng* Roots," *Planta Medica*, 50: 434–436, 1984.

24. Hikino, H., Y. Oshima, Y. Suzuki, and C. Komo, "Isolation and hypoglycemic activity of panaxans F, G and H, glycans of *Panax ginseng* roots," *Shoyakugaku Zasshi*, 39: 331–333, 1985.

25. Oshima, Y., C. Konno, and H. Hikino, "Antidiabetes drugs. XIV. Isolation and hypoglycemic activity of panaxans I, J, K and L, Glycans of *Panax ginseng* roots," *J. Ethnopharmacol.*, 14: 255–259, 1985.

26. Konno, C., M. Murakami, Y. Oshima, and H. Hikino, "Validity of the Oriental medicines. CVI. Antidiabetes drugs. 19. Isolation and hypoglycemic activity of panaxans Q, R, S, T and U, glycans of *Panax ginseng* roots," *J. Ethnopharmacol*, 14: 69–74, 1985.

27. Chang, H. M. and P. P. H. Butt, *Pharmacology and Applications of Chinese Materia Medica*, Vol. 1., World Scientific, Singapore, 1986.

Section VI

The Healing Power of Ginseng

15 Traditional Chinese Medicine Records of Medical Benefits of Ginseng and Siberian Ginseng

Traditional Chinese medicine (TCM) describes ginseng in the Shen-nong-Ben-cao Jing (100 BC) as being sweet and a little cold. It mainly supplements the five viscera. It quiets the essence spirit, settles the ethereal and corporeal souls, checks fright palpitations, eliminates evil qi, brightens the eyes, opens the heart, and sharpens the wits. Protracted consumption may make the body light and prolong life.

To appreciate fully the ancient description of ginseng, here is an elaboration of the statements:

- *Ginseng is sweet*: Anyone who tastes a ginseng root will find it quite bitter. The freshly picked root is sweeter (has a somewhat pleasant taste compared to many other herbs), but, more importantly, the designation of the root as sweet is partly based on the idea that sweet is the underlying inherent taste within the herb that reflects its actions. Like other sweet herbs, it is believed that ginseng will supplement the spleen, calm irritation, and nourish the body. Later Chinese texts often mention the bitter taste as well.
- *A little cold*: The action of the herb is relatively mild, in contrast to a cold herb, but its nature is still like that of a cold herb, able to alleviate heat syndromes. The combination of sweet and cold together has the effect of calming nervous agitation: the sweetness alleviates irritation, and the coldness calms the internal fire that agitates the human spirit.
- *Supplement the five viscera*: Ginseng does something more than most of the sweet herbs; it benefits not only the spleen, but also the other four systems of the body: liver, kidney, heart, and lung. One implication is that ginseng greatly improves the ability of the spleen to draw nutrients out of food and distribute them to the other organs. It serves as a nutritive aid but does not provide nutrients.
- *Quiets the essence spirit*: The essence spirit can be broadly interpreted as referring to the mind. Thus, ginseng quiets the mind. By taking ginseng, excessive mental chattering calms down.

- *Settles the ethereal and corporeal souls*: The ethereal soul (*hun*) and corporeal soul (*po*) refer to fundamental forces within the body. The ethereal soul is believed to reside in the liver and to be responsible for dreams; the unsettled *hun* causes one to have disturbing dreams, even nightmares. The corporeal soul is said to reside in the lungs, and to be responsible for maintaining the integrity of the physical body. Persons who develop lifelong degenerative diseases are said to suffer from scattering of the corporeal soul, often the result of being frightened. It could be said that ginseng calms the distressed mind while strengthening the body.
- *Checks fright palpitations*: When a person is frightened, he/she experiences an irregular heartbeat and palpitations. Anxiety attacks and panic attacks correspond roughly to fright palpitations. Ginseng helps calm the heart (the resting place of the spirit) so that it does not overly react to external stimuli or to internal mental worries: equanimity is restored.
- *Eliminates evil qi*: Evil qi refers to influences from the environment that cause diseases. Herbs that nourish the viscera, like ginseng, are usually not attributed the ability to expel the evil that is causing disease; more often, such herbs are said to protect the body from evil qi (evil cannot get into the strong and well-nourished body to cause disease) or to help the body recover its strength once the evil has been eliminated. Ginseng can be taken when a disease is present to help cure it by eliminating the evil qi. Some later authorities disagreed with this view, suggesting that ginseng had only tonic properties and should not be taken while evil qi was still present for fear of enticing the evil to stay within the body.
- *Brightens the eyes, open the heart, and sharpens the wits*: The eyes are the windows to the mind; the heart is the residence of the mind, and the wits are the expression of the mind. This section says that by taking ginseng one's mind will not become dull. If the heart becomes closed, if the mind becomes overwhelmed with thoughts, if the spirit is clouded and the eyes therefore dim, then a person's fundamental nature will be prevented from attaining its ultimate expression: the person will be timid, unhappy, even depressed. When the heart opens and the mind quiets, the true nature will be expressed, and the person will display sure purpose, will, and courage, and be able to accomplish great things.
- *Protracted consumption may make the body light and prolong life*: This phrase is included in reference to the intensive efforts undertaken by Taoists pursuing immortality during the Han dynasty period. They believed that one could shed the physical body and float into the heavens as an immortal being. Most of this transformative (alchemical) process was accomplished with minerals, such as cinnabar (mercury sulfide), which slowly poisoned the Taoists' bodies due to prolonged exposure. One consequence was that they lost weight: at the time, their shrinking flesh was taken not as an indication of poisoning, but as a sign that they were shedding their early body to leave only the heavenly body. This section does not indicate that ginseng can be used as a weight loss herb for the obese, nor does it even suggest that one can live longer on this earth by taking the herb regularly; it refers specifically to the Taoist concept of transformation to an ethereal immortal.

Ginseng's value as a universal tonic, which promotes and maintains good health, particularly for those who are physically and mentally troubled, aged men and women, has long been recognized in the Orient, and is slowly gaining recognition in the Western world. For quite a long time Chinese medical practice has claimed that ginseng has the power to restore vigor, maintain health, and erase the debilities of old age. Numerous scientific experiments in laboratories and clinical trials in humans have helped to establish that ginseng preparations are indeed capable of stimulating the functions and regulating the malfunctions of the organs, the central nervous system, the cardiovascular system, the endocrine glands, and the metabolism.[1] In addition, ginseng is unique in medicinal properties in that it remarkably increases an organism's resistance to harmful stresses arising from various origins, thus maintaining the normal functions of the organism.[2]

For thousands of years the Chinese as well as other Orientals used ginseng, primarily, for *tonic* purposes. Chinese doctors believe that most ailments are due to the debility of the body or the loss of balance or harmony of the organism. Accordingly, for mental and physical debility ginseng is the best remedy. On the other hand, Soviet scientists have confirmed that ginseng is a harmless *adaptogen*. The real truth about ginseng, the evidence on "how ginseng works," and "what ginseng can do for you" are discussed in the following chapters.

ADAPTOGENIC POWER OF GINSENG

The distinguished pharmacologist I. I. Brekhman and his associates of the Institute of Biologically Active Substances, Siberian Branch of the Academy of Sciences of the Russian Federation, Vladivostok, conducted systemic biological research on Oriental herb drugs, particularly ginseng and Siberian ginseng, for more than 25 years. Evidence from laboratory and clinical investigations indicated that the basic effect of ginseng's activity is its capacity to increase the nonspecific resistance of the organism toward various stresses.[3–5] This characteristic medical claim is called *adaptogenic activity*. The concept of adaptogenic activity represents a recent innovation in Western medicine, a new milestone in modern therapeutics.

ADAPTOGEN

According to Brekhman and Dardymov, *Panax ginseng* and several other species of the Araliaceae family provide the medical properties of adaptogens.[4,5] The word *adaptogen* means "a substance causing a state of nonspecifically increased resistance of the organism to stresses of various origin." Ginseng is one of the most useful adapteogens.[5] The concept of "a state of nonspecifically increased resistance (SNIR)" of the organism was originally developed by N. V. Lazarev, who found that 2-benzyl-benzimidazol (bendazol) was effective for the treatment of damage to various regions of the nervous system, and for increasing nonspecific resistance of the organism to adverse stresses.[6] As a matter of fact, the discovery of the adaptogenic activity of bendazol and its analogous drugs was already known in well-established Oriental herb medicine.

REQUIREMENTS TO BE AN ADAPTOGEN

What types of medicinal agents have the virtue of adaptogens? The requirements for a remedy to be an adaptogen, according to Brekhman, are that it should (1) b innocuous or safe; (2) have antistress activities; and (3) possess normalizing and protective effects. According to published research reports, ginseng and Siberian ginseng (eleuthero) meet all of the fundamental requirements necessary to be classified as adaptogens. One of the most important indices of a drug's adaptogenic action is its capacity to increase a human's physical resistance toward adverse stresses and maintain the body in homeostasis.[5] Thus, ginseng overcomes diseases by a mechanism of building general vitality and resistance and strengthening the normal functions of the organism. Brekhman proclaimed that ginseng, without any doubt, is an "adaptogen," if not a "panacea."[5]

The body has a homeostatic mechanism to maintain a constant equilibrium, and medicinal substances that help maintain homeostasis were named "adaptogens"; Brekhman explained that adaptogenic activity is the essence of the tonic effect of ginseng.

HARMLESS MANROOT

When we use a drug, we want to know its purity, side effects, and relative toxicity and safety. The laboratory LD_{50} index, or 50% animal lethal (death) dosage, is the method most frequently used in pharmacology and toxicology to indicate the toxicity of the drug.

After repeated tests, it has been proved that ginseng root, its extracts, and its chemically purified constituents (ginseng saponins) are harmless. The well-known ginseng pharmacologists Brekhman,[5] Yonekawa,[7] Kitagawa and Iwaki,[8] Hong,[9] and Kaku et al.[10] have repeatedly stressed the fact that ginseng has very low toxicity in comparison with a majority of commonly used *official* remedies. Yonekawa[7] found that the lethal dose of ginseng saponin, ginsenin, was 2–3 g/kg in mice, while Kitagawa and Iwaki[8] found that the LD_{50} of ginseng ethereal extract in mice was 5 g/kg. Brekhman[5] reported that the LD_{50} of *Panax ginseng* root was 10–30 g/kg, while the LD_{50} for the isolated pure ginseng saponins, panaxosides, given orally were 1.4 g/kg in mice. The oral toxicity of ginseng root is about 10–20 times less than that of its pure saponins. Even for the pure ginseng saponins, the LD_{50} of ginsenoside Rg_1 is 1.25 g/kg given intraperitoneally to mice.[10] The toxicity of ginseng saponin is similar or lower than the toxicity of most commonly used *official* drugs. For example, the oral LD_{50} value (in rats) of aspirin is 1.75 g/kg; caffeine, 0.2 g/kg: and sulfaquanidine (an intestinal antibacterial agent), 1 g/kg. Thus, ginseng is an *innocuous* and perhaps even a safer remedy in comparison with those officially accepted as "drugs."

STRESS AND ANTISTRESS ACTIVITY OF GINSENG

WHAT IS STRESS?

Stress, in a broad sense, is mental or physical strain, pressure, tension, or unhappiness. Frustration and suffering are also stressful conditions. Adverse stresses can be man-made or due to the environment. The most common chemical and physical sources of stress are drugs, chemicals, fumes, polluted water and air, high or low pressures, extreme temperatures, and radiation. Biological causes of stress may be bacteria, toxins, foreign sera, viruses, and tumor tissue. In addition to these physical, environmental, and biological stresses, there are also mental stresses of a socioeconomic nature.

Stress can make you miserable or ill. In extreme cases, it can kill you. But how do you respond to stress? For a long time, doctors have told us to avoid stress, but Hans Selye, considered the world's leading authority on stress, says that the key to handling stress is to work at something at which you can win. You can overcome the effects of stress by following a code of behavior based on natural laws.[11]

STRESS-RELATED DISEASES

Evidence is growing that stress can be one of the components of any disease. The presence of stressful environmental conditions is one of the major factors that contribute to an individual's susceptibility to disease, as pointed out by Robert Ader, Professor of Psychiatry and Psychology at the University of Rochester School of Medicine and Dentistry[12]:

> "Every illness from the common cold to cancer, has a psychologic component," said Dr. Samuel Silverman, Associate Clinical Professor of Psychiatry at Harvard Medical School. Some physicians, including Dr. Silverman, still think that the seven well-defined diseases [peptic ulcer, hypertension, hyperthyroidism, rheumatoid arthritis, ulcerative colitis, neurodermatitis, and asthma], are *psychosomatic* diseases, that is, the physical disorder is caused by the emotional state of the patient.[12]

HOW YOUR BODY HANDLES STRESS

Laboratory studies on animals have shown that the adrenal gland in particular is the prime reactor to stress. For example, when rats are exposed to various types of stress such as environmental temperature change, pressure change, injecting foreign protein or toxic material, all the rats had their adrenals enlarged, while the lymph nodes, thymus, spleen, and gastric and duodenal tracts had shrunk. The initial response of the body to any kind of stress is alarm, followed by the stage of resistance.

Selye, the world-famous endocrinologist and director of the American Academy of Stress Disorders, published his original concept of reactions toward stress, the "nonspecific biological stress syndrome" describing the pituitary-adrenal axis that releases powerful corticosteroids in the body's precautionary girding for "fight or flight" when responding to danger or startle.[12]

TYPICAL STRESS EXAMPLES

Stress, such as meeting the annual income tax return deadline of April 15, may lead to serious physiological changes, which may even result in cardiovascular disorders. Meyer Friedman and Ray Roseman of Mount Zion Hospital and Medical Center followed the blood changes of 18 tax accountants between January and June.

As a rule, the accountants increased their workload to 70-hour weeks against the April target date, then slacked off to leisurely 30-hour weeks. In the course of this experiment, the group's serum cholesterol averaged 206–217 mg% during the off-season winter months of work, rose sharply in March, and peaked at a mean 323 mg% by mid-April. Then by the first week in June, it declined to 206 mg%. Blood clotting times kept pace, from 8.1 minutes in February, dropping to a tense, taut 5 minutes on April 15, and by June increased again to 8.8 minutes. This economical stress on the accountants caused high blood cholesterol, one of the effects of stress in persons prone to cardiovascular disorders.[13]

Both Sidney Cobb of Brown University and Robert Rose of Boston University School of Medicine agree that because of the round-the-clock and day-in and day-out stress of split-second decisions and actions, air-traffic controllers have a higher-than-average incidence of hypertension, peptic ulcers, and diabetes.[14] Hypertension has long been thought to be a psychosomatic disease—the disease caused by stress. Elevated blood pressure has been specifically associated with emergency situations, with prolonged combat duty, and with job termination.[14]

How would unemployment affect one's health? In 1975, the unemployment rate was headed for 10% nationwide, but in an economic disaster area such as Detroit, the automobile city, 15%–20% of the work force was idle. Gordon Deshler, medical director of Detroit's Metropolitan Health plan, said that "The truth of the matter is that Detroit is probably one of the most stressful areas in the whole world. We have large numbers of people who require continuing support with medication for pain, anxiety, for stress, and so forth. It really is pretty much a way of life."[12]

Antistress Effect of Ginseng

Quite a number of interesting laboratory investigations dealing with the protective or antistress effect of ginseng have been reported in the medical literature. It is of great interest and importance that some properties of ginseng are demonstrable only in stressed, injured, and impaired animals. In other words, normal animals or animals kept at normal conditions usually show little or no physiological change after being fed ginseng. This is because animals usually retain their homeostasis. Therefore, in order to evaluate the true antistress effect of ginseng, various types of experiments were designed using animals under severe stress or animals with previously induced stresses of a physical, chemical, or biological nature.

Physical stresses, such as extremely cold and hot temperatures, have been used to evaluate how the animals react, and what role ginseng plays in the reaction to stress. Quantitative measurements of the vitamin C (ascorbic acid) content in the adrenal gland of the test animals and their body temperature have been used as indictors to evaluate stress. This is because under extreme temperatures, either cold or hot, the adrenal vitamin C of the animals usually falls to a very low level.

It seems that the earliest laboratory studies on temperature stress started some 20 years ago. In 1958, two Chinese physiologists, Sung and Chi of the Chinese Academy of Medical Sciences, Beijing, announced their fundamental research discovery on the protective effect of ginseng against temperature stresses.[15] In their study, white male rats were divided into groups: one group ate a regular diet, while the other group ate the same diet supplemented with 5% ginseng powder. After 3 weeks, some of the rats from both groups were put into a 78°C–90°C chamber for 5–6 minutes or put into a chamber at −2°C for 1 hour before being returned to room temperature. In another experiment, each rat received 2.4 mL of 50% ginseng extract 1 hour before the temperature stress, while rats fed with water served as controls. After temperature stress, all of the rats were killed, and the vitamin C contents of their adrenal glands were determined. It was found that the vitamin C (ascorbic acid) content of the adrenal glands of rats subjected to either hot or temperature stress was significantly depleted. However, depletion of the adrenal vitamin C was less in rats rerated with ginseng powder in their diet or with ginseng extract 1 hour before the stress. Also, the rats in the control group subjected to high-temperature stress either appeared unable to move or showed chronic convulsion, and did not regain their normality until 20–60 minutes after their removal from the hot chamber. Most of the rats fed with ginseng appeared to be normal immediately after their removal from the hot chamber, and none of them showed convulsion. Their work clearly indicated that ginseng was capable of increasing the nonspecific resistance of the organism to temperature stress. The authors suggested that the administration of ginseng extract may interfere with the nervous control of the pituitary gland or that ginseng has adrenocorticotropic hormone (ACTH)–like action.[15,16]

Similar antistress effects observed with ginseng could in some cases also be obtained with cortisol or cortisone. Thus, the question has been raised whether the antistress effects of ginseng are actually mediated through the adrenal cortex or through the brain pituitary-adrenal cortex system. Studies in the area of endocrinology are very difficult. Many times, Brekhman has said that the intravenous injection of ginseng extract to a dying cat, after it has lost its breath and heartbeat can often revive it. Chang and Kao injected a ginseng extract to dogs dying as a result of bleeding or suffocation, and brought the dogs to life. All of these indicate that the action of ginseng is definitely related to the adrenal system or the pituitary gland of the brain or both.[17]

In 1964, the effect of ginseng against temperature stress was investigated by Tsung, Chen, and Tang of Kirin Medical Institute.[18] In their experiments, instead of measuring the adrenal vitamin C level, they explored the possible mechanism of the organism's antistress activity by observing how the animal survived. Ninety-two male white mice were treated by intraperitoneal injection of a 50% aqueous extract of ginseng, 10 mL/kg, to each mouse; while those injected with the same amount of physiological saline served as controls. In each experiment, six mice (three experimental and three control) were put into a 45°C–47°C chamber 30 minutes after the injection or into a 2°C chamber 20 minutes after the injection. The mice were allowed to remain in the chamber until half of the mice were dead. The results showed that out of 21 mice in each group, 14 mice in the control group and 7 in the experimental group were dead under heat stress, while among the 25 mice in each group kept in the cold chamber, 18 in the control group and 6 in the experimental group were dead.

In a second series of experiments, the adrenals of the mice were taken out, and by a similar procedure, they were put in the cold and hot temperature chambers. The results showed that out of 17 mice in each group, 7 in the control group and

10 in the experimental group were dead in the hot chamber, while out of 20 mice in each group, 10 in the control group and 9 in the experimental group were dead under cold stress. Their experiments clearly confirmed that ginseng strikingly raised the animals' ability to tolerate temperature stress, and elongated their survival under stress. However, this antistress activity was abolished after the removal of the adrenals. The authors then suggested that the mechanism of the antistress activity of the organism induced after the injection of ginseng might be associated with the hypophysis (pituitary gland) adrenal system.[18]

Research on the antistress activity of ginseng was also pursued in Bulgaria. In 1963, Petkov and Staneva-Stoicheva of Postgraduate Medical Institute, Sofia, reported their work on the effect of ginseng on the adrenal cortex.[19] In their experiments, white rates were fed with 20% alcoholic extract of ginseng by means of a stomach tube at a daily dose of 2 mL/kg for 17 days, while a second group of rats, fed with the same volume of water, served as controls. In one series of their experiments, they found that the eosinophil count in the peripheral blood of the experimental rats was 26% lower than at the beginning of the experiment. The control animals had only a slight reduction in the number of eosinophil cells. The average weight of the adrenals of the experimental rats was increased by about 13% over that of the controls. Also there was a slight decrease in the vitamin C content of the adrenal cortex, but the corticosteroid content in the urine was found to be 60% higher in the experimental animals than in the controls. Another series of experiments was performed under hot stress by immersing one hind leg of the rat in hot water (70°C) for 1 minute. After this stress, the eosinophil count of the control animals had a 41% rise after 2 hours and a 45% rise after 4 hours of the stress, while the rats under ginseng treatment, after the same intervals, showed a 12% and a 9% fall of the eosinophil count. During the subsequent days three rats in the control group died, while all the rats in the experimental group survived. In a third series of experiments, all of the rats had one of their adrenal glands removed (adrenalectomy). After this unilateral adrenalectomy, the remaining adrenal gland of the ginseng-treated rats, in each case, weighed much more than the adrenal gland that had been removed at the beginning of the experiment. The corresponding increase in weight of the adrenal gland of the rats in the control group was insignificant. Accordingly, the hypertrophy effect (enlargement) of the adrenal gland of the experimental rats indicates the pharmacological action of ginseng. All of these results, according to Petkov, confirmed that ginseng extract had a stimulating effect on the adrenal cortex, and this effect may be of neurogenic origin. Ginseng increased the reactivity of the cerebrocortical cells, thus facilitating the adaptation of adrenal cortical function to the needs of the organism under stress. The anti-inflammatory and antiexudative effects of ginseng, as had been reported by others, may also be due to the stimulating effect of ginseng on glycocorticoid hormone production.[19]

After the studies of Sung and Chi, Tsung and associates, and Petkov, Wang of the Chinese Medicine Research Institute in Kirin further reported that a 20% alcoholic extract of ginseng leaves and stems had a stimulating effect on the activities of the pituitary-adrenal cortex system. From different pharmacological experiments, Wang confirmed that the activity of ginseng was not directly on the adrenal cortex per se but rather on the cerebral level, that is, the pituitary gland of the brain.[1]

Antistress effects have been observed not only with ginseng ethanol extract but also with a number of isolated ginseng glycosides. Ginsenoside Rb_1 or ginsenoside Re was administered to rats for 10 days. From the eighth day, the rats were exposed to heat (45°C, for 30 minutes) for three consecutive days. It was observed that the decrease of adrenal ascorbic acid content due to stress was inhibited by the application of isolated pure glycosides, indicating that glycoside also normalizes the insufficiency of adrenal glands induced by stress. Here again in normal animals, adrenal ascorbic acid content did not change significantly after treatment with pure glycosides.

In recent years, studies on the antistress effect of ginseng were performed in much greater detail by many investigators in South Korea. For example, in 1964, B. I. Kim reported that an alcoholic extract of ginseng, as does hydrocortisone, prolonged the survival time of the adrenalectomized mice when exposed to cold.[21] In 1965, C. Kim reported that the total serum protein, hemoglobin, and the red blood cells of mice were markedly decreased when exposed to a cold environment. However, the ginseng-treated mice showed an increase in the serum hemoglobin and the red blood cell counts.[22] In 1965, G. C. Kim reported the influence of ginseng on the rectal temperature of mice in a cold environment in comparison with cortisone and chloropromazine.[23] In his experiments, 1,680 mice were divided into four groups; a normal group without restraint (movement of all four legs were uninhibited), a normal group with restraint, adrenalectomized mice without restraint, and adrenalectomized mice with restraint. These mice were exposed to a cold temperature of 0°C (50 minutes per day) for 1, 5, and 10 days. The adrenalectomy (both sides) was performed 3 days before the cold stress. It was found that the mice treated with ginseng (an alcoholic extract at a daily dose of 10 mg/kg) and treated with cortisone (at a dose of 10 mg/kg/day) repressed the drop of rectal temperature of all groups, and the suppression effect was more apparent with the mice in the restraint groups. Both ginseng and cortisone exerted a greater effect on the normal mice than the adrenalectomized mice in preventing the rectal temperature drop. It is surprising, however, after the administration of chlorpromazine (a dose level of 9.5 mg/kg/day), that both the normal and adrenalectomized mice with restraint showed a further drop in the rectal temperature.[23]

Similar studies on the influence of ginseng on the rectal temperature of rats exposed to cold was also reported by Yoon and Kim in 1971. Two hundred adult male albino rats were divided into ginseng and saline groups. For 8 or 15 consecutive days, the rats received 5 mL/kg of either a ginseng extract or a physiological saline. The significant findings from their studies were that without exposure to cold environment, there was no significant difference in rectal temperature between the ginseng and the saline groups, regardless of whether the rats were intact (without any surgery), hypophysectomized, adrenalectomized, thyroidectomized, or thyroid adrenalectomized; the rectal temperature of the ginseng-treated rats after the different types of surgery as mentioned dropped little in comparison with those in the saline group; also ginseng facilitated the earlier recovery from abnormal to normal rectal temperature than saline in the rats under cold stress.[24]

Hu and associates reported their series of studies on the effect of ginseng against temperature stresses. Their findings were essentially similar to those reported by others, indicating that ginseng exerted little or no influence on the adrenal vitamin C content of animals under normal temperature conditions. However, the administration of ginseng did accelerate the recovery from, and helped in the maintenance of, the body temperature under cold or heat stress conditions.[25–29]

Only recently W. B. Kim, H. R. Kim, and H. Y. Chung found that after the administration of ACTH, the adrenal vitamin C content decreased significantly as it does with temperature stress. However, animals, both mature and immature, under the influence of ginseng showed a protective effect and mitigated the fall in the adrenal vitamin C content.[30,31]

Ever since the announcement of ginseng's adaptogenic power by Brekhman, scientists all over the world tried to prove and also disprove its marvelous action. Amirov and Abdulova reported that the increase in the resistance of the body to toxic substances can be achieved not only by gradual adaptation to toxic substances but also by the administration of ginseng. As an adaptogen, ginseng has polytropic action. It improves the resistance of the body toward stresses by any mechanism that may act upon the endocrine system, metabolic processes, and may be joined together with neurohumoral mechanisms, thus stimulating the amazing defensive or adaptive activities of the body.[32]

The tendency to get sick and also the ill effects of exposure to toxic chemicals such as morphine, cocaine, strychnine, and curare poisons were much less in ginseng-fed animals than in control animals.[33–35] Animals under ginseng treatment also showed relatively longer survival time during periods of starvation,[36] or when exposed to low pressure and extremely cold temperature after adrenal gland removal.[37] The tolerance of ginseng-treated mice to the stress of increased gravitational force was greater than that of the control group,[38,39] another incident showing the antistress effect of ginseng.

NORMALIZING EFFECT OF GINSENG

The third and most important attribute of an adaptogen is its normalizing effect on wounds and fast recovery from various illnesses. In experimental investigations with animals, it has been established that ginseng is effective against chemically (alloxan) induced diabetes, radiation sickness, experimental emotional disorders, mental strain and nervous exhaustion, hypotension, and certain forms of hypertension and cancer.[5] This broad-spectrum antistress effect of ginseng is due to its normalizing effect. The "normalizing effect" of ginseng or of any other adaptogenic drug is a new dimension in medicine and a novel approach in therapeutics. The following are a few examples to illustrate the marvelous normalizing effect of ginseng.

Liver damage caused by carbon tetrachloride or irradiation was prevented by the administration of ginseng extract to animals.[40] Ginseng also prevented the development of fever induced by intravenous injection of typhoid and a paratyphoid vaccine in rabbits. When mice were injected with trypanosomes (a genus of flagellate protozoan), it was found that the life span of the ginseng-treated animals was significantly prolonged compared to that of the controls.[40] Rabbits inoculated with *Shigella paradysenteriae* W. and given ginseng extract suffered only from a drastically increased number of leukocytes (white blood cells) in the blood, whereas the control animals developed diseased white blood cells (leukopenia).[41]

Ginseng has certain anti-inflammatory activities. It appears to be capable of lessening the swelling at the ankle joints due to the injection of egg-white or dextran.[42]

A noticeable anti-inflammatory effect was obtained in rabbits having an inflammation of the ear-shell caused by freezing.

In general, laboratory experimental work revealed that in animals exposed to pathological environmental conditions, the ginseng-fed animals showed quick adjustment with respect to the adverse conditions. This is the typical normalizing of the organism under the influence of ginseng. In addition to the medical defensive power, the wonder root is further capable of aiding in recovery from damage such as abnormal enlargement (hypertrophy) and contraction (atrophy) of the adrenal and thyroid glands caused by toxic chemicals; reducing the blood sugar level in animals having very high blood sugar; or increasing the blood sugar level in animals having very low blood sugar. The normalization effect was also attained in red blood cell diseases as well as white blood cell diseases.[43,20]

After learning all the facts about ginseng's mysterious adaptogenic power, we know no chemical drug today that is comparable to ginseng. No wonder the manroot deserves the title of "panacea," or "elixir of life," and has been scientifically called "adaptogen." In my opinion, to maintain homeostasis and good health, we all need this mysterious adaptogen against all known and unknown harmful stresses and distresses.

SIBERIAN GINSENG'S MEDICAL BENEFITS

Both *Panax Ginseng* and Siberian Ginseng are well-known herbal medicines in China for centuries. The dried root and rhizome of *Acanthopanax senticosus* of the family Araliaceae is Siberian ginseng. It grows in abundance in northeastern China, particularly in the Heilongjiang province, as well as Inner Mongolia, North Korea, and Siberia. The varieties of Siberian ginseng *Eleutherococcus henry* and *E. senticosus* are known to most Westerners. The Chinese variety of Siberian ginseng, or *wu jia shen*, is not well known; it is a tonic herb for *Qi* and an effective antifatigue herb. It is also recommended for rheumatism, bone or tendon pain, general debility, and weakened immunity. Siberian ginseng is an adaptogenic and an effective, antiaging medicinal herb.[44,45]

The best known medical benefits of Siberian ginseng are the adaptogenic effect: Siberian ginseng, or *wu jia shen* (*ci wu jia*), has even a stronger adaptogenic activity than *Panax ginseng*. An adaptogen is a substance that causes "a state of nonspecific increased resistance" (SNIR) to adverse stresses of various origins.[46] An effective adaptogen must (1) be innocuous and safe, (2) have antistress activity, and (3) possess normalizing and protective properties. According to published results, ginseng, Siberian ginseng, and Chinese schisandra fruits meet all the fundamental requirements necessary to be classified as adaptogens.[46]

Siberian ginseng invigorates *Qi* and nourishes the spleen; it is used for weakness, fatigue, lassitude, anorexia, insomnia, sleep with dreams, forgetfulness, and palpitations as a result of weakness of the spleen and *Qi* deficiency. Siberian ginseng can be used alone in a decoction, in a tincture, in an extract such as *Wu Jia Shen Gao* (R-7), or in a combination with schisandra fruit and other tonic herbs.[47]

The *Qi* tonic effect has healing properties. The antifatigue and antistress properties in Siberian ginseng are stronger than in *Panax ginseng*. Also, a noticeable sedative effect on the central nervous system may occur.[47] The herb also stimulates immune resistance and can be taken in convalescence to aid recovery from chronic illness. As a general tonic, Siberian ginseng helps both to prevent infection and to maintain well-being. Siberian ginseng is also used in the treatment for impotence.

Siberian ginseng nourishes the kidneys and tranquilizes the spirit: it is used for treating male sexual disorders, impotence, spermatorrhea, and aches and pains in the loins and knees due to kidney deficiency or mental disturbance. Siberian ginseng is prescribed with processed rehmannia root, eucommia bark, dodder seed, and rosa cherokee fruit[47] to stimulate adrenal and sex hormone production, it is also a gonadotrophic.[47]

The German Commission E, on the basis of studies conducted concluded that eleutherosides were an effective tonic.[48]

The glycosides of Siberian ginseng appear to act on the adrenal glands, helping in the prevention of adrenal hypertrophy and excess corticosteroid production in response to stress. The eleutherosides additionally help reduce the exhaustion phase of the stress response and help the adrenals return to normal function faster. As a result, Eleuthero root has a beneficial effect on the heart and circulation. It has been shown to increase energy and stamina, and to help the body resist viral infections, environmental toxins, radiation, and chemotherapy.

Siberian ginseng or *Wu jia shen* is a sedative and is helpful in patients suffering from sleep difficulties. The root is used to treat a wide spectrum of ailments in the nervous, cardiovascular, and endocrine systems. It helps to adjust blood sugar levels and is used to treat sexual debility.[46,49]

Siberian ginseng expels pathogenic wind and eliminates dampness: it is also used to treat arthritis, rheumatic arthralgia, and numbness of the limbs. It can be used alone in a decoction or in combination with millettia (*ji xue teng*), dried (chaenomeles) papaya fruit, and clematis root (*wei ling xian*).[47]

Siberian ginseng regulates endocrine secretions, adrenal cortex, and blood sugar levels.[47] It also appears to produce moderate reductions in blood sugar and blood cholesterol levels and modest improvements in memory and concentration. Siberian ginseng may also have mild estrogenic effects. In laboratory studies, various chemicals found in eleuthero have also shown antiviral and anticancer properties, but these effects have not been well studied in humans.

Siberian ginseng increases the immune activity of the body, and it has a protective effect against radiation and toxic chemicals that cause loss of white blood cells. Evidence is also mounting that Siberian ginseng enhances and supports the immune response. Siberian ginseng may be useful as a preventive measure during cold and flu season. Recent evidence also suggests that Siberian ginseng may prove valuable in the long-term management of various diseases of the immune system, including HIV infection, chronic fatigue syndrome, and autoimmune illnesses such as lupus.

Siberian ginseng is able to strengthen the immune system and improve resistance: it is used to treat white blood cell loss in cancer patients caused by the side effects of chemotherapy or intoxication of noxious chemicals (such as pyridoxine or benzene fumes). Siberian ginseng extract, or *Wu Jia Shen Gao* (R-7), can be used to strengthen immunity and boost detoxification.[47]

Dong reported that Siberian ginseng has an anti-inflammatory effect and pain relieving effect. It also has analgesic and anti-inflammatory effects.[49]

IMMUNE SYSTEM

Evidence is also mounting that Siberian ginseng enhances and supports the immune response. Siberian ginseng may be useful as a preventive measure during cold and flu season. Recent evidence also suggests that Siberian ginseng may prove valuable in the long-term management of various diseases of the immune system, including HIV infection, chronic fatigue syndrome, and autoimmune illnesses such as lupus.

In perhaps the most convincing study carried out so far, B. Bohn and coworkers in Heidelberg, West Germany, looked at immune parameters in 18 individuals in a randomized, double-blind fashion for a total of 4 weeks. The subjects in this study had venous blood drawn both before and after *Eleutherococcus senticosus* administration, and the samples were analyzed by flow cytometry, which counted absolute numbers of immune cells present in their blood.

Overall, the *E. senticosus* group showed an absolute increase in all immune cells measured. Total T-cell numbers advanced by 78%, T-helper/inducer cells went up by 80%, cytotoxic T cells by 67%, and NK cells by 30%, compared to the control group. The B lymphocytes, which are cells that produce antibodies against infectious organisms, expanded by 22% in the *E. senticosus* subjects, compared to controls. Most importantly, no side effects were noted in the *E. senticosus* subjects up to 5 months after *E. senticosus* administration ended.

The researchers stated: "We conclude from our data that *Eleutherococcus senticosus* exerts a strong immunomodulatory effect in healthy normal subject." The Bohn study has caused drug companies to spend millions of dollars in an effort to get *E. senticosus* approved as a drug by the U.S. Food and Drug Administration.

The increases in T, B, and NK cells in people given *E. senticosus* suggest that it could be very useful in alleviating immune suppression associated with strenuous exercise. In addition, one might speculate about a positive effect of *E. senticosus* in the very early stages of HIV (AIDS-virus) infection. In an HIV-infected patient, *E. senticosus* might prevent or retard the spread of the virus, thanks to the synergistic positive actions of elevated numbers of both helper and cytotoxic T cells.

Supporting these findings, *E. senticosus* is now used in the support of cancer patients undergoing radiation and chemotherapy, especially in Germany. Studies have shown that when administered to patients, *E. senticosus* drastically reduces the side effects of radiation and chemotherapy (e.g., nausea, weakness, fatigue, dizziness, and loss of appetite). Other research with cancer patients has linked *E. senticosus* with improved healing and recovery times, increased weight gain, and improved immune cell counts. In Russia, the administration of *E. senticosus* to cancer patients seemed to permit larger than normal doses of drugs utilized in chemotherapy, thus speeding treatment periods.

How does *E. senticosus* actually spur the immune system to greater activity? At present, there is no consensus. Some researchers believe that *E. senticosus* induces increased interferon biosynthesis (interferon is a powerful chemical that boosts immune system activity), while others believe that polysaccharides (long-chain sugar molecules) naturally found in *E. senticosus* stimulate the activity of special white blood cells called *macrophages*. These macrophages play a number of roles in the immune system, including the breakdown of infected cells and the stimulation of other immune cells. However, the polysaccharides are probably "nonspecific" immune stimulants, which means that their effectiveness fades fairly quickly and that they must be administered continuously or at regular intervals in order to produce a positive effect.

ATHLETES AND ANTIBIOTICS

Why should athletes try to stimulate their own immune systems, rather than rely on antibiotics and other remedies to control infections? Obviously, prevention of infection can promote more consistent, high-quality training and lower the risk of missed competitions. In addition, many microorganisms are now resistant to many of the commonly used antibiotics. That means that an infection picked up during heavy training may be more difficult to shake off than ever before.

Some of the more notable antibiotic-resistant organisms include *Streptococcus pyogenes*, which causes "strep throat", upper respiratory infections, and is reported to be resistant to both penicillin and chloramphenicol. Another common bacterial species, *Haemophilus influenzae*, which produces both ear and upper-respiratory tract infections, is now resistant to a variety of antibiotics, including chloramphenicol, ampicillin, and tetracycline. *Staphylococcus aureus*, which causes "staph infections" of the skin, especially around surgical wounds, is resistant to erythromycin, tetracycline, and the ß-lactam antibiotics. Finally, certain strains of *Escherichia coli*, which have caused deaths in recent incidents when customers of restaurants have consumed contaminated or poorly cooked meat, are resistant to a variety of different drugs.

Investigators in the United States recently completed a pilot study in which *E. senticosus* extract was given to AIDS patients in hopes of improving their immune system functioning and overall survivability. The results were very promising, and so a four-city, randomized, double-blind clinical trial will be carried out with *E. senticosus*.

Extracts of *E. senticosus* appear to have the ability to prevent immune suppression in vigorously training athletes and may limit the risk of infection. By boosting recovery following hard workouts, *E. senticosus* may also downgrade athletes' chances of overtraining.

The ability of Siberian Ginseng to increase stamina and endurance led Soviet Olympic athletes to use it to enhance their training. Explorers, divers, sailors, and miners used eleuthero to prevent stress-related illness. After the Chernobyl accident, many Siberian citizens were given eleuthero to counteract the effects of radiation.

REFERENCES

1. Wang, P. H., *Acta. Pharm. Sinica*, 12: 477, 1965.
2. Brekhman, I. I., *Medgiz*, Leningrad: 182, 1957.
3. Brekhman, I. I., *Med. Sci. Serv.*, 4: 17, 1957.
4. Brekhman, I. I. and I. V. Dardymov, *Lloyda*, 32: 46, 1969.
5. Brekhman, I. I. and I. V. Dardymov, *Ann. Rev. Pharmacol.*, 9: 419, 1969.

6. Lazarev, N. V., in *7th All-Union Congress of Physiol. Biochem. Pharmacol.*, Medgiz, Moscow, 1947, 579 pp.
7. Yonekawa, M., *Keijo J. Med.*, 6: 785, 1926.
8. Kitagawa, H. and R. Iwaki, *Nippn Yakurigaku Zasshi (Folia Pharmacol., Japan)*, 59: 348, 1963.
9. Hong., S. A. and H. Y. Cho, in *Korean Ginseng Science Symposium*, Korean Society of Pharmacognosy, Seoul, 1974, pp. 113–139.
10. Kaku, T. et al., *Arzneim-Forsch.*, 25: 539, 1975.
11. *Industry Week*, January 30, 1975.
12. Leff, D. N., *Med. World News*, 16: 74, March 24, 1975.
13. *Med. World News*, 16: 103, February 24, 1975.
14. Cobb, S. and R. M. Rose, *J. Am. Med. Assoc.*, 224: 489, 1973.
15. Sung, C. Y. and H. C. Chi, *Acta Physiol Sinica*, 22: 155, 1958.
16. Liu, K. T. and C. Y. Sung., *Acta Physiol Sinica*, 25: 129, 1962.
17. Chang, T. H. and T. H. Kao, *J. Chin. Med. Assoc.*, 44: 1040, 1958.
18. Tsung, J. Y., C. Cheng, and S. Tang, *Acta Physiol. Sinica*, 27: 324, 1964.
19. Petkov, W. and D. Staneva-Stoicheva, in *Proc. 2nd. International Pharmacology Meeting, Prague, August 20–23*, 1963, Pergamon Press, NY, 1965, pp. 39–45.
20. Brekhman, I. I. and O. I. Kirillov, *Rast. Resursy.*, 4: 1, 1968.
21. Kim, B. I., *Choongang Uihak (Korean Med. J.)*, 8: 107, 1963.
22. Kim, C., *Katorik Taehak Uihakpu, Nonumnjip*, 8: 251, 1964.
23. Kim. C. C., *Katorik Taehak Uihakpu, Non Mumjit*, 9: 29, 1965.
24. Yoon, H. S. and C. Kim, *Katorik Taehak Uihakpu, Non Mumjit*, 21: 25, 1971.
25. Hu, C. Y. and C. Kim, *Katorik Taehak Uihakpu, Non Mumjit*, 12: 49, 1967.
26. Kim, C. C., J. K. Kim, and C. Y. Hu, *Katorik Taehak Uihakpu, Non Mumjit*, 10: 455, 1966.
27. Hu, C. Y., C. C. Kim, and J. K. Kim, *Choosin Uihak*, 10: 73, 1967.
28. Kim, C. C., J. K. Kim, and C. Y. Hu, *Choosin Uihak*, 10: 57, 1967.
29. Kim, C., C. C. Kim, M. S. Kim, C. Y. Hu, and J. S. Rhe, *Lloydia*, 33: 43, 1970.
30. Kim, H. R. and W. B. Kim, *Choesin Uihak*, 15: 87, 1972.
31. Kim, W. B. and H. Y. Chung, *Choosin Uihak*, 15: 83, 1972.
32. Amirov, R. V. and E. B. Abdulova, *Uchenye Zapskii Azerbaidzhanskogo Zast. Usoversh Urachei*, 1: 3, 1966.
33. Kim, C. -c. et al., *Choesin Uihak*, 10: 57, 1967.
34. Oshima, Y., *J. Chosen Med. Assoc.*, 19: 539, 1929.
35. Golikov, P. P., *Inf. A Study Ginseng and Other Drugs the Soviet Far East*, 7: 17, 1966.
36. Hong, S. A., *Choesin Uihak*, 15: 87, 1972.
37. Kim, C.-C. and Roh, H.- K., *Katorik Taehak, Uihakpu. Non-Munjip*, 8: 265, 1964.
38. Kim, C. C., *Korean Med. J.*, 11: 51, 1966.
39. Choi, Y. C., *Seoul Uidae Chapchi*, 13: 1, 1972.
40. Chang, P.-H., *Yao Hsueh Hsueh Pao (Acta Pharm. Sinica)*, 13: 106, 1966.
41. Brekhaman, I. I. and N. K. Fruentov, *Farmakol. I Toksikol.*, 19(suppl.): 37, 1956.
42. Zhou, J. H., F. Y. Fu, and H. P. Lei, "Proceedings of the Second International Pharmacological Meeting," in *Pharmacology of Oriental Plant*, Vol. 7, K. K. Chen and B. Mukerji, Eds., Pergamon Press, NY, 1965, p. 17.
43. Pichurina, R. A., *Dissertacija, Tomskiy Medicinskiy. Institut.*, Tomsk, 1963, p. 106.
44. Hou, J. P., *The Myth and Truth About Ginseng*, South Brunswick, NJ: AS Barnes and Co., 1978.
45. Jiang Su New Medical College, *Encyclopedia of Chinese Materia Medica*, Shanghai Science and Technology Press, Shanghai, 1977.
46. Brekhman, I. I. and I. V. Drdymo, *Annual Review of Pharmacology*, 9: 419–430, 1969.

47. Wang, J. H., Ed., *Xin Bian Chang Yong Zhong Yao Shou Ce* [Manual of Commonly Used Chinese Medicinal Herbs], Jin Dun Press, Beijing, 1994.
48. Robbers, J. E. and V. Tyler, *Tyler's Herbs of Choice*, The Haworth Press, Binghamton, NY, 1999.
49. Dong, K. S., X. Q. Wang, and Y. F. Dong, *Xian Dai Lin Chuang Zhong Yao Xue* [Contemporary Clinical Chinese Materia Medica], Zhong Guo Zhong Yi Yao Press, Beijing, 1998.

The section on Siberian Ginseng's Medical Benefits is taken from *"The Healing Power of Chinese Herbs and Medicinal Recipes"* by Joseph P. Hou and Jin, Y. Y., The Haworth Integrative Press, NY, 2005. (p. 94–96).

16 What Ginseng Can Do For You

SAFE TONIC AND THE ANTIFATIGUE EFFECT OF GINSENG

FATIGUE

Fatigue is the common symptom of weariness that occurs after sustained or intensive physical or mental strain. Fatigue causes irritability, decreased ability to concentrate, a tendency to be upset by trivialities, and impairment of the ability to rationalize. Most often, psychiatric fatigue, such as combat fatigue during a period of war, is a very common symptom of distress, uneasiness, or emotional disorders. Common symptoms associated with these disorders are headache, chest pain, very rapid beating of the heart, along with difficulty in breathing. Certain illnesses such as anemia, chronic infection, malnutrition, metabolic disease, and endocrine gland disorders may cause chronic fatigue.

TONIC EFFECT OF GINSENG

Repeated laboratory tests have shown that the active principles of ginseng root indeed are able to prevent fatigue and to increase the physical performance of animals, including man. "The increase in strength and efficiency after a single dose is called *stimulant action*, while the improved vitality after prolonged administration of ginseng is called *tonic effect*," according to Brekhman.[1]

The stimulant effect of ginseng was first tested in humans by the Chinese 2,000 years ago. The test consisted of a person with a piece of ginseng root in his mouth walking for *5 li* while another walked with his mouth empty, and if the person with ginseng in his mouth did not feel tired or out of breath, the root in his mouth was genuine ginseng (the tonic effect). The other person must be out of breath. This two-man walk test was the first trial in history to identify if the ginseng was genuine or an adulterant.

The first modern method of detecting the *tonic* and *antifatigue* effects of ginseng in animals was a swimming test using white mice of similar weight.[2–5] In order to make the results more obvious and the testing period of swimming shorter, a 1 g weight was put on the tails of the mice. Results from hundreds of tests showed that the average swimming time after oral administration or injection of 0.15 mL of ginseng extract, was increased to about 60%–100% over controls.[1,3,5]

The antifatigue effect of ginseng is more pronounced if ginseng has been taken continuously for a longer time. Again, by using the swimming test with mice, Soviet scientists conducted experiments that continued for 2 months, during which time the mice had to swim once every 5 days. For the first 10 days on which they swam, no ginseng preparation was given. From the 11th day to the 40th day, half of the mice received injections in an amount of 0.1 mL per mouse (20 g of body weight) of a 10% aqueous liquid extract once every other day. Mice in control received 2% alcohol solution. The swimming was started 20 minutes after the injection.

At the end of the 2-month test, the average swimming time of the mice in control was about 47–61 minutes, while those under the influence of ginseng was about 96–117 minutes, which is about an 80% improvement. It was further noted that in the 20-mouse groups, 8 in the control group and only 4 in the ginseng-treated group were dead as a result of complete physical exhaustion.[6]

A second method is to measure the increased duration of running on an endless rope.[7] The device consists of four closed, vertical boxes $7 \times 7 \times 25$ cm. Through the center of each box, a descending rope is passed. The rope is put into motion by an electric motor at a velocity of about 6 m/min. The floors of the boxes are electrified to about 25 volts. The time to complete fatigue is that time when the mice can no longer run but remain on the floor despite the current being on. In this particular test, the mice under the influence of ginseng gave a significantly better performance than the control.

In order to test the potency of the ginseng preparations, the amount of the preparation that increases the work duration by about 33% is called one stimulant unit of action, denoted as SUA_{33}.[6] This particular pharmacological testing method of ginseng preparations was developed by Brekhman at Vladivostok. It was shown that the active principles, ginseng saponin glycosides isolated by Soviet chemists from the Chinese ginseng root, gave much more pronounced stimulant action than the crude extract. This was proved by the swimming and rope-running tests.[8] One SUA_{33} equals 0.101 mL/20 g of body weight of 15% solution of ginseng extract, or 0.151 mg of ginseng saponin glycoside, panaxoside C, powder. A number of ginseng saponins and aglycones (obtained after acid hydrolysis of saponins) isolated from the ginseng root are about 100–1,000 times more potent than the crude extract. Among the five ginseng saponins isolated by the Soviet chemists, panaxosides A and C (panaxatriol group) are more potent than panaxosides D, E, and F (panaxadiol group). Also, panaxoside C, with four sugar molecules, is twice as potent as panaxoside A, which has only three sugar molecules. The aglycone, panaxatriol, possesses a relatively higher activity than panaxadiol. Of the panaxadiols, panaxoside D is the most potent. Panaxoside E and panaxoside F are a pentoxide and a hexoside, respectively, and are less active than panaxoside D. The potencies of panaxosides are about 700–6,600 SUA_{33}, while the potencies of the aglycones are about 2,000–5,000 SUA_{33} in comparison with ginseng extracts, which are only 50–70 SUA_{33}.[3] Furthermore, the activities of ginseng extracts from natural ginseng were virtually the same as those cultivated ginseng roots.

The Japanese workers, without delay, repeated the swimming tests of mice conducted initially by the Soviet investigators; however, a slightly different approach was adopted. In the tests, four different types of ginseng extracts were used: ether extract, alcohol extract, water total extract (without previously extracting with organic solvents), and water extract (after extraction with ether). Five mice in a group were to swim at 32°C until exhaustion, and the swimming performance of each mouse was recorded. As shown in Table 16.1, with the mice in the control groups the duration of swimming was from 34 (shortest) to 194 (longest) seconds, while with the mice

TABLE 16.1
Swimming Performance in Mice with and without Ginseng Extract Treatment

		Ether Extract	Alcohol Extract	Total Water Extract	Control Group		
					I	II	III
Dose in mg/kg	Test A	200	100	100			
		55–123	52–164	80–198	34–96	40–96	46–194
Duration of swimming (in seconds)	Test B	74–280	131–577	104–242	40–116	41–101	52–180

Note: The swimming tests were performed at 32°C, five mice in a group. The durations of swimming (from the shortest to the longest) were recorded and are shown. *A* represents the swimming tests conducted 3 days after oral administration of the ginseng extract; *B* represents the swimming tests conducted 1 hour after administration of the ginseng extract.

after treatment with ginseng extract, the swimming performance was significantly increased. It was especially noted that the mice treated with alcohol extract, ether extract, and water extract, respectively, showed the swimming duration, in each case, to be improved by from 60% to 200% in comparison to both the shortest and the longest swimming records of the controls. Particularly noteworthy was the comparison between the performance observed 1 hour after treatment with ginseng extract to that conducted 3 days after the oral treatment with ginseng extract. Thus, 1 hour after treatment with ginseng the mice's performances were much better than the controls.[5]

The antifatigue effect of ginseng was also tested using special commercially available ginseng preparations of GI Pharmaton and Geriatric Pharmaton. The swimming time of rats after administration of these preparations was significantly prolonged over controls. Geriatric Pharmaton gave a much more pronounced effect than the GI Pharmaton preparation.[9]

The most recent mice swimming studies were performed by K. H. Rueckert of Pharmaton, Ltd., Lugano, Switzerland.[10] In his first experiment, 450 mice were used. These mice, both male and female of similar body weight of about 20 g, were divided into nine groups of 50 mice each. The swimming performance of these mice was measured after the mice had been treated with Pharmaton's ginseng extract in the diet. Two dose levels of the ginseng extract were used, that is, 3 mg/kg/day or 0.06 mg/20 g (mouse body weight), and 30 mg/kg/day or 0.6 mg/20 g. Each of these dose levels was administered continuously to groups of mice for periods of 14, 21, and 28 days. At each time interval, the swimming performance of the ginseng-treated mice was tested and compared with that of the untreated control group. In the swimming test, the mice were placed in a water bath at 18°C and allowed to swim until complete exhaustion. The swimming time of each mouse was recorded. After the first swim, the mice were allowed to dry out in a warm airstream. Each mouse was then retested after a rest-drying period of 1 hour. When the swimming tests were completed, the average or mean swimming time in each of the two tests of the two dose groups was computed. The results showed that the mice treated with the 0.06 mg/day dose for

14 days improved little, while the mice treated with 0.6 mg/day had better swimming performance than the control group. The improvement was about 12%–20%. After 21 days of treatment with ginseng extract, the swimming performance of 0.6 mg/day group, for example, was about 27% better at the first test and about 20% better at the second test than the control group. After 28 days of ginseng extract treatment, the swimming performance was significantly improved by about 48% at the first test, and 38% at the second test over the control group; the mice treated at a lower dose level, 0.06 mg/day, also showed improvement in swimming performance but not as much as those at the higher ginseng dose level of 0.6 mg/day. The data on swimming tests thus collected are listed in Table 16.2. The table may give you a better understanding of the improved swimming performance of the ginseng-treated mice over the controls.

Using a similar procedure, by using 1,000 mice, Rueckert repeated the swimming study again in 1975. These mice were numbered and divided into several groups of 50 mice (25 female and 25 male of similar body weight of about 20 g) in each group. For each of the treated groups there was a separate control group of 50 mice, which were kept under the same conditions. They were allowed to swim at $18 \pm 0.1°C$ until complete exhaustion, and the swimming time of each mouse was registered. After the

TABLE 16.2
Comparison of Swimming Performance of Mice at $18 \pm 0.1°C^a$

Group	Test	Mean Swimming Time (Seconds)	Performance, Percentage (%) Improvement/Control
1. Control	First	476 ± 92	—
	Second	545 ± 121	—
2. 0.06 mg, 14 days	First	502 ± 97	5.4%
	Second	594 ± 133	9.0%
3. 0.6 mg, 14 days	First	582 ± 100	22.0%
	Second	610 ± 122	12.0%
4. Control	First	496 ± 86	—
	Second	536 ± 83	—
5. 0.06 mg, 21 days	First	592 ± 104	19.5%
	Second	654 ± 140	22.0%
6. 0.6 mg, 21 days	First	630 ± 133	27.0%
	Second	645 ± 137	20.2%
7. Control	First	466 ± 82	—
	Second	518 ± 96	—
8. 0.06 mg, 28 days	First	701 ± 130	51.8%
	Second	719 ± 134	38.8%
9. 0.6 mg, 28 days	First	668 ± 139	47.9%
	Second	717 ± 132	38.4%

[a] Each mouse was treated (except the control group) with ginseng extract. Geriatric-Pharmaton, at levels of 0.06 or 0.6 mg per mouse for 14, 21, and 28 days, respectively.

first test, a dry-resting period of 1 hour was allowed, and then the mice were tested again (second test). Before the swimming experiment was conducted, the mice were treated (except the control group) with ginseng extract G-115 at 0.06 mg/day/mouse for, respectively, 14, 21, and 28 days. The controls were treated with placebo (water) daily. The results showed that mice treated with ginseng extract for 14 days had about 5.1% increase in swimming performance at the first test and about 9.0% at the second test over the controls.

Figure 16.1 shows the increase in swimming performance in mice treated with ginseng extract, Geriatric Pharmaton G 115. The percent of increase is obtained by *subtracting* the mean control time from the swimming time of mice treated with ginseng, *divided* by the mean swimming time of the control, *multiplied* by 100.

After 21 days of treatment, the swimming improvements were about 20% at the first test and about 22% better at the second test over the controls. After 28 days of treatment, however, the improvement was even more pronounced, about 52% at the first test and about 39% at the second test over the control group. These numbers are plotted in the figure. The figure depicts the percentage increase in swimming performance of mice after treatment with 0.06 mg/mouse/day of ginseng preparation (ginseng extract F-115) for 14, 21, and 28 days over the control mice. As a matter of fact, the curve clearly indicates that a much better improvement of swimming performance, called *duration*, was obtained after a relatively longer ginseng treatment.[11]

In a separate study, mice treated with two other ginseng extract preparations of two foreign companies, X2 and S-II, for 28 days, also showed some improvement in swimming performance over the controls. It seems that ginseng extract is indeed capable of improving the physical activities of animals, and the improvement could be

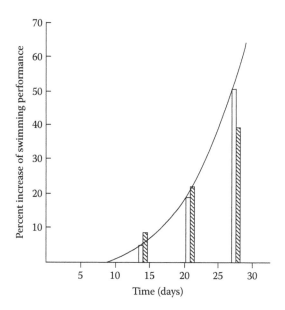

FIGURE 16.1   Percentage increase of swimming performance.

reached at a relatively low dose. The biggest problem in this type of experiment would be the purity and strength of the ginseng preparation. In order to obtain dependable and reproducible results, a standardized product has to be used.[11]

The antifatigue activity of ginseng extract was also investigated by the well-known Japanese pharmacologists Saito, Nabata, and Takagi of the University of Tokyo. In their studies, six different testing methods, for example, exploratory movement, hole cross, rotating rod, sliding angle, spring balance, and rectal temperature tests were used. They evaluated the effect on different fractions of ginseng root extract on recovery from exhaustion. Four hours of oscillation exercise were used to exhaust the mice. Aqueous extracts of ginseng were injected intraperitoneally immediately following the exercise. It was observed that a water extract of ginseng significantly accelerated the recovery of exploratory movement and elevated rectal temperature. The antifatigue effect of a ginseng saponin, ginsenoside Rg_1, and the lipophilic fraction of ginseng extract was more pronounced and more obvious in every test than that of other fractions isolated by the Japanese investigation.[12,13]

The swimming and running data clearly suggest that ginseng's antifatigue effect in mice may also be applicable to other animals and to humans for athletics or sports.[14] Quite interesting results were shown with horse racing in Japan and Korea back in the early twentieth century (Chapter 4). A study with ginseng extract (Geriatric Pharmaton, G115) in 42 sportsmen was conducted with a physical education unit in Liverpool, England. These 42 students had to undergo a training program with continuous medical checkups. This particular study showed that the sportsmen's stroke volume and cardiac output, reaction time, and respiratory quotient were significantly increased; in other words, the athletic performance, in general, was improved.[11] Having such a promising effect, the number of athletes from the Communist countries treated with ginseng tonic before attending the Olympic games is unknown.

Not too long ago, vitamin E, a modern chemical agent, had been claimed to prevent, among other things, ulcers, baldness, arthritis, diabetes, liver dysfunction, sexual impotence, and aging. The same vitamin has been added to cosmetics and used as a deodorant. Other claims have suggested that large doses of vitamin E can improve athletic performance[15]; however, in controlled trials, the E vitamin did not enhance the endurance of competitive swimmers.[16–18] The likelihood of serious adverse effects of vitamin E from such self-medication appears to be very low. Nevertheless, diarrhea and intestinal cramps have been reported with daily dose of 3200 IU of vitamin E.[18]

The antifatigue mechanism observed with ginseng and how it improves physical performance have been the subject of argument between Soviet and Chinese investigators. The Russian scientists, particularly Brekhman's group, are saying that ginseng basically is a "stimulant," and it has a marked stimulant effect.[19] On the contrary, based on their experiments, the Chinese investigators, Tsung et al., of Kirin Medical Institute, proclaimed that the antifatigue activity of ginseng has a "sedative" instead of a "stimulant" effect.[20] The antifatigue action of ginseng was shown by the fact that the treated animals (rats) decreased their oxygen consumption, thus reducing the unnecessary tension and muscle exercise and fatigue, and their survival in a low-pressure (190/Hg) chamber was also prolonged. This was confirmed with three experiments. In a mice swimming test, the swimming duration time of mice in the control group was 60 ± 10 minutes; the time for the ginseng-treated group,

88 ± 8 minutes (about 50% better than the control group); while the time for the caffeine-treated group was 47 ± 3 minutes. In another experiment, the mice were kept in reduced-pressure chambers. None of the ginseng-treated mice were dead, 10 of 20 mice injected with sodium bromide were dead, and all of the 20 mice injected with caffeine were dead. A third experiment was the oxygen-consumption tests. It was found that the least oxygen was consumed after the injection of ginseng, in comparison with the amount consumed after injection of sodium bromide and chlorpromazine, and the caffeine group showed the highest oxygen consumption.

Scicenkov studied the effect of ginseng on the visual performance of healthy young men. He found that, after a single administration of the liquid extracts of ginseng roots, the process of dark adaptation was accelerated, its final level was increased, and visual acuity also increased substantially. This may also be considered to be a tonic effect.[21]

It is an undisputable fact that ginseng can be considered as a tonic, a "safe" tonic. As Brekhman proclaimed about ginseng, "It possesses the stimulant action of other commonly used stimulants, but its safety is unique. It decreases a person's fatigue and boosts a person's working capacity as well as efficiency." In clinical pharmacological tests, the tonic effect of ginseng on humans becomes apparent after both single and repeated doses. The tonic effect was observed not only during the treatment, but also 3–4 months later. The general tonic effects observed during the course of treatment with ginseng are good mood, good appetite, good sleep, and absence of depressed state. Ginseng's tonic effect on animal organisms (including man) has been proven.[6]

Brekhman, the Russian proponent of using ginseng as adaptogens, later described the tonic effect of ginseng as a stimulant action. Chinese researchers insisted that ginseng functioned as a mild sedative and calming agent; the ability to overcome fatigue, for example, was described as the result of having less stress on the body rather than by causing an overt stimulation. One peculiar effort to resolve the differences in these viewpoints was made by a Japanese researcher, Hiroshi Saito, who suggested that ginseng did both: some of the ginsenosides, particularly the Rg series, were stimulating in nature, while others, particularly the Rb series, were calming in nature. He and coworker Yien-mei Lee, pointed out that "We noticed that multiple pharmacodynamic activities of ginseng originated from various ingredients and there are many pharmacologically antagonistic actions in ginseng." How these apparently competing effects of some active components could explain away the difference of opinion about ginseng effects when ginseng was used as a whole preparation and not subdivided was never made clear by any subsequent authors who used the underlying concept in the attempt to cover up incompatible claims.

As an example of the confusion generated by those making claims for ginseng's effect on energy, it has been suggested that Asian ginseng is "stimulating," but that American ginseng is "calming." One origin of this concept may be traced in the Chinese view regarding yin and yang. In China, it has been said that Chinese ginseng has the ability to tonify qi and invigorate yang (corresponding roughly to metabolism and movement), while American ginseng has the ability to tonify qi and nourish yin (corresponding roughly to control and calming). However, this analysis comes about from a peculiar historical situation rather than an inherent difference in the herbs. Chinese ginseng comes from the far north of China; American ginseng was always imported through Hong Kong, in the far south. From this experience, the Chinese had viewed their own ginseng as a northern product and the American ginseng as a southern product. In fact, American ginseng is mainly from the northern part of the United States and Canada.

CAN GINSENG PREVENT CANCER?

Modern medical science still does not know how the change from a normal to an abnormal cell takes place. This is the main riddle of cancer. Certainly you are aware that noxious chemical fumes, polluted air and water, excessive cigarette smoking, excessive exposure to radiation, certain food dyes, food preservatives, food additives, and certain pesticides are possible carcinogens—agents that cause cancer.[22] With the advancement of the chemical and drug industries, new chemicals and drugs are placed on the market every day. More new synthetic drugs and chemicals introduced on the market means more potential carcinogens.

In the United States, the cancer incidence is increasing steadily. In 1973, 37,000 persons died from colon and rectal cancer; 14,700 from stomach cancer; and 6,400 from esophageal cancer.[23] The American Cancer Society estimated the death toll in 2000 to be 552,200, almost 12 times the 46,181 American victims of combat in Vietnam, and 59,000 more than the 295,867 Americans who died on the battlefields during World War II.[24]

Enough evidence leads us to believe that excessive smoking causes lung cancer. The risk of dying from lung cancer for those who have smoked heavily for years is about 20 times greater than for nonsmokers.[25]

Breast cancer is also increasing and occurring in a younger age group than before. At least 1 in every 20 women in the United Kingdom was found to be suffering from breast cancer.[26]

Cancer may also be induced by psychological or nervous stress, particularly in women. "There is some casual link between cancer and the feeling of hopelessness, helplessness, despair, and depression," said James P. Henry, Professor of Physiology at the University of Southern California School of Medicine. Henry has paid a great deal of attention to studying why his laboratory female animals get breast tumors. He thinks the cause is despair, not rage, as when female mice are separated from their young.[27]

Aging is an important determinant of the incidence of cancer. Regardless of what the cancer may be, about 1 man in 14 of those 60 years of age or older, as compared to 1 in 700 of those 25 or younger will develop cancer.[25]

Scientists have been desperately searching for anticancer drugs for years. Those available to date that act on gastrointestinal tract malignant tumors are mitomycin, carmustine (BCNU), lomustine (CCNU), fluorouracil, fluorometholone, thiopepa, and nitrosourea. These potent chemotherapeutic agents have demonstrated some tumor inhabiting effect, yet none of them have shown outstanding success in the treatment of cancer.[28]

Russian scientists have devoted enormous time to ginseng research in the area of cancer prevention, if not treatment, in the past 25 years. In experimental oncology, ginseng preparations were capable of inhibiting urethane-induced adenomas of the lung, 6-methyl-thiouracil–induced tumors of the thyroid gland, and indole-induced myeloid leukemia in animals.[29–31] Cells of Ehrlich ascitic tumor were introduced intravenously into mice, and ginseng exerted an effect of decreasing the transplantability and the size of tumorous foci.[32] Evidence was also obtained in the decrease of formation of spontaneous tumors of the mammary gland[33,34] and spontaneous leukemia in mice.[35] Ether and alcoholic extracts of ginseng had a much greater effect than the aqueous extract in inhibiting malignant tumors in the secretory glands. However, ginseng had no effect on leukemia-1210 tumors.[36] With these preliminary findings, the use of ginseng or its components in the prevention of cancer may be possible in the future.

Ginsenosides, the essential oil and the polysaccharides of ginseng, showed antitumor activity. Among the individual saponins, ginsenosides Rb_1, Rb_3, Rd, Re, Rf, Rg_1, Rg_2, and Rh_1 were tumor inhibitory. Ginseng polysaccharides inhibited the growth of Ehrlich ascites carcinoma and sarcoma S_{180} in mice. Ginseng essential oil inhibited the stomach cancer cell line SGC-823 *in vitro*. When ginsenosides were added to the culture medium of rat Morris hepatoma cells, morphological changes toward normal cells were observed in the cultured hepatoma cells.

Ginseng also exhibited *anticarcinogenic* activity. Oral administration of red ginseng extract attenuated the carcinogenicity of urethane, N-methyl-N′-nitro-N′-nitrosoguanidine and aflatoxin in mice and rats.

Furthermore, ginseng could increase the activity and reduce the toxicity of other antitumor agents. An emulsion preparation containing the mixture of ginseng extract and the antitumor agent tegafur increased the survival of mice bearing leukemia P_{388} more effectively than tegafur used alone. Ginseng polysaccharides also produced a synergistic antitumor effect with cyclophosphamide.

A recent report from Tang Center for Herbal Medicine Research and Department of Anesthesia Critical Care, University of Chicago, reported the anticancer effect of American ginseng.

American ginseng can potentially be used for cancer treatment and chemotherapy-induced side-effect management. In *in vitro* studies, American ginseng was found to inhibit the growth of breast cancer cells. After steaming treatment of American ginseng, its antiproliferative effects on cancer cells were improved significantly, possibly due to the altered ginsenoside profile. Antiproliferative effects of representative constituents were also evaluated, showing that ginsenoside Rg has a positive effect. Steamed American ginseng inhibited the colorectal cancer growth both *in vitro* and *in vivo*, which might be achieved through all-cycle arrest and induced apoptosis in the cells.

The cellular and molecular targets of ginsenosides against cancer have also been studied. It appears that several molecular mechanisms exist and collectively converge on various signaling pathways. These pathways include regulation of cell cycle, induction of apoptosis, inhibition of angiogenesis, prohibition of invasion, and reduction of inflammatory response. A series of cell cycle proteins, apoptosis-related proteins, growth factors, protein kinases, and transcription factors are affected by ginsenosides. For example, Rh2 and Rg3 inhibit cancer cell proliferation by inducing gene and protein expression of the cell cycle regulatory protein p21, thus arresting tumor cell cycle progression by inducing cancer cell apoptosis through activation of caspase-3 protease via a bel-2-insensitive pathway and by sensitizing multidrug-resistant tumor cells to chemotherapy. To characterize further downstream genes targeted by ginseng saponins such as Rg3 in a human cancer cell line, the gene expression profiling was assayed, showing that the most affected pathway was the Ephrin receptor pathway.

The most commonly used cancer chemotherapies are limited by severe side effects and dose-limiting toxicity. The drug-related adverse events not only worsen patients' quality of life, but can also lead to refusal to continue a potentially curative chemotherapy. American ginseng and ginsenoside Re attenuated cisplatin-induced nausea and vomiting in a rat model without affecting its anticancer properties in human cancer cells.

IMPOTENCE AND APHRODISIAC ACTIVITY

Impotence is the most common sexual disability after 45 years of age. Sexual failure is a serious problem; it can cause serious distress or uneasiness,[36] and it may precipitate an emotional crisis in a man that may threaten his life and his marriage.[37]

Factors causing male impotence include the following:

- *Psychogenic*: 90% is due to psychological factors
- *Organic diseases*: Debilitating diseases
- *Fatigue*
- *Intoxication*: Alcohol, drugs, and marijuana

Most sexual disorders including male impotence are mainly psychological in origin. However, even if you experience impotence, you may well still be able to attain an erection and function sexually under certain circumstances, for example, with a partner who knows how to make love, or with the help of a sexual stimulant remedy.[37,38] Most important of all, you must forget all your troubles and concentrate on love-making. Impotence may also be a consequence of organic disease. About half of all male diabetics become impotent before the age of 60.[38] Fatigue is another important factor leading to impotency. The husband, exhausted by his day's work, loses interest in love when he gets home; he may be overdedicated to his job or his profession; his wife thinks he is married to his job, not to her. Besides physical fatigue, mental fatigue due to job difficulties, economic crisis, and depression may also be factors causing male impotence.

Statistically, heavy alcohol drinking is a common factor in impotence.[39] Marijuana, on the other hand, interferes with production of sexual hormones, in certain cases suppressing the male sex hormone, testosterone, to levels that could result in impotence and even infertility.[40]

There is no actual age limit for having or not having sex. There are 30-year-old men who are inactive, and there are 90-year-old men who still have active sex lives. Those leading a vigorous sex life in younger years tend to remain active longer. Good health, overall satisfaction with life, and the availability of a sanctioned sexual partner help maintain sexual activity's significant role in the life of the elder person.[41,42]

At age 68, about 78% of men still engage regularly in sexual activity. There are studies that have reported that sex around the age of 90 or 100 is nothing unusual if a person is healthy and physically fit.[43]

APHRODISIAC REMEDIES

All through history, people have searched for some means of revitalizing weakening sexual potency. All kinds of treatments from all over the world have been tried, without notable success. Today, scientists finally may be on the trail of an authentic aphrodisiac. Several substances have shown attractive properties and may be useful as sexual stimulants for those who are impotent, particularly the aged.[44]

Many popular magazines carry advertisements advocating the use of oral preparations, sprays, and ointments for the genitalia that will increase sexual potency. The ingredients of these sexually stimulating preparations usually include vitamin E, whose role is quite unproved, together with varying mixtures of minerals and trace elements, carbohydrates, and proteins. Few of the preparations have any proven pharmacologic effect.[45]

Spanish fly and yohimbine are some of the most well-known traditional aphrodisiacs.[44] Spanish fly is also called cantharides, or blistering fly. It is obtained from iridescent beetles found in southern France and Spain and is an extremely harsh irritant. When swallowed in liquid form, it burns the mouth mucous membranes. According to Kent, the irritating effect of Spanish fly can cause dilation of the blood vessels of the genital organs, leading to erection of the penis or congestion of the labia. Rather than being pleasurable, such physical arousal is unpleasant and dangerous. It causes gastroenteritis and nephritis. In fact, a large dose of Spanish fly can be fatal. Yohimbine is a powder made from the inner bark of African yohimbe tree. It acts similarly on the blood vessels, but its aphrodisiac effect has been questionable.[44]

Parachlorophenylalanine (PCPA) is a chemical that can produce an aphrodisiac effect in rats. However, PCPA is a very dangerous chemical that causes convulsions and other negative reactions. As a result, it has not been tested in humans.[44]

Some people who tout marijuana as having an aphrodisiac capacity contend that they enjoy sexual relations more when under the influence of this perception-distorting drug. However, it has been proven that the sexual-stimulating effect of marijuana is similar to that of alcohol.[44]

Levodopa has been used extensively in humans and has produced significant hypersexual behavior in men when it was administered for the treatment of Parkinson disease.[46] There is controversy about the use of androgen (male sex hormone), however, and in most cases it is found useless in the treatment of impotence.[47]

GINSENG'S APHRODISIAC ACTION

The great popularity of ginseng is probably based on its aphrodisiac activity. The reputed power of ginseng as an aphrodisiac has attracted the attention of millions of people.

Perhaps Eiberhof was the first Western scientist who tested the active principles from ginseng (panaquilon) in patients and found it exerts some aphrodisiac action.[48] Yona Cawa reported that ginseng was active as an aphrodisiac and general mild stimulant to the sympathomimetic system.[49] Min found that ginsenin produced a tail erection effect in mice that may be related to aphrodisiac properties.[50] Panax acid obtained from an ether extract of ginseng also produced a tail erection in mice.[51] The extracts from *P. ginseng* exhibited greater activity than that from American ginseng in producing the tail erection effect in animals.[50]

The book *Ninjinshi (History of Ginseng)* written by T. Imamura, recorded that in the 1930s many Japanese and Korean hospitals used ginseng extracts in the treatment of impotence and some female sexual disorders.[52]

Some recent scientific investigators have provided more evidence for the gonadotropic or aphrodisiac action of ginseng. One study showed that ginseng has estrogenic properties.[53] In experiments with animals, a gonadotropic effect of ginseng was established as manifested by causing infantile male and female mice to reach puberty earlier than they would have otherwise. The gonadotropic activities of ginseng saponins, panaxosides A, C, D, E, and F, exerted a similar, but stronger effect than that of crude ginseng extract. However, Kit[54] and Wang[55] reported that ginseng does not possess any male or female sex hormone activity, based on the fact that administration of ginseng preparations could not restore the castrated male and ovariectomized female animals to normal. However, ginseng contains certain principles that do stimulate and promote an animal's sexual maturity.

Ginseng was given orally to (normal) rats in a daily dose of 0.1–0.2 g for 61 days, or the rats were injected with 10 mg of methanol extract of ginseng for 8 days. There was no remarkable change in growth of either female or male rats, nor was any change observed in the sexual cycle between experimental and control rats.[56]

Whether or not the active principles of ginseng root contain any sex hormone activity has been puzzling for a long time, and only a few reports are available to date. Female hormone-like activity was observed in one study after a methanol extract of ginseng was given to female mice.[57] By a thin-layer chromatographic technique, estrogen-type (female hormone) components of estrone, estradiol, and estriol were isolated from an oil-soluble fraction of ginseng extract.[58] This is the first time that ginseng was found to contain sex hormones or sex hormone-like substances.

Using a biochemical approach, it was found that an alcoholic extract of ginseng increased the testicular RNA and DNA levels of rats. Oura and associates observed in a series of investigations that ginseng extracts, particularly fraction number four, increased the synthesis of DNA and protein in test tubes. It is assumed that the synthesis of RNA and protein may be equally promoted in the testes.[59–61]

Brekhman administered Korean ginseng to 44 patients who had showed refractory impotence to all medication for 2–5 years, and observed complete recoveries of 21 patients and considerable improvements of 11 patients.

Popov and others administered ginseng to 27 impotent patients, of whom 15 completely recovered and nine showed considerable improvement. So ginseng proved to be an excellent agent for revitalization therapy.

Impotence and decreased libido are extremely frequent in diabetic patients. Kakiwoochi reported that *Panax ginseng* was shown to be effective in the management of the complications in diabetic patients, and made considerable improvement in 80% of impotent patients and 60% of patients with decreased libido.

In another recent experiment using mesenteric mast cells as indicators, it was found that ginseng preparations have the ability to increase the number of mast cells. In the castrated male rat, ginseng functions like testosterone by decreasing the deterioration of the mast cells.[61]

Recent clinical studies in humans in Japan and in the Soviet Union show that ginseng extract was quite effective toward several forms of impotence, particularly the premature ejaculation-type impotence and nervous-type impotence. It is not effective for psychic-type impotence.[55] The aphrodisiac activity of ginseng seemed to be proven correct. As a sexual tonic, ginseng is probably the safest aphrodisiac agent in comparison with all other chemicals speculated around the world.

A man may take ginseng to treat erectile dysfunction (ED). A 2002 Korean study revealed that 50% of men who took ginseng noticed an improvement in their symptoms. In addition, research published in the *British Journal of Clinical Pharmacology* provides "evidence for the effectiveness of red ginseng in the treatment of ED".[129]

Another report shows that ginseng has been used for both men and women.

If you take ginseng for long enough, often even just once, there is absolutely no denying that it is a powerful sexual tonic, increasing libido, performance, and pleasure for both men and women. For men, ginseng's benefits include boosting libido, increasing pleasure, and treating erectile dysfunction, by increasing nitric oxide production similar to prescription medications like Viagra.[62] *Panax ginseng* has similar benefits for women, most notably increasing libido, pleasure, and sensation, also in part due to its oxygenating effects throughout the body.

QUIETING YOUR SPIRIT

The central nervous system (CNS) is the most highly organized of all the systems of the body. It may be considered as being superimposed on all other organs for the purpose of coordinating their diverse functions toward a harmonious functioning of the organism as a whole. Because of its complex structure, delicate ramifications, and supreme importance to the body, the CNS is extremely sensitive to all drugs. How does ginseng affect the CNS? This problem has been the number-one research project, and a dozen of the top pharmacologists in the world have investigated it.

The pharmacologic properties of ginseng claimed in the earliest Chinese Materia Medica books are

quieting the spirit, establishing the soul, allaying fear, brightening the eyes, benefiting the understanding.

This is a beautiful and factual description of ginseng's activities on the CNS. How did the Chinese doctors know all about this even 2,000 years ago? Only after 50 years of scientific research has it been proven that the activities of ginseng on the CNS are its principle pharmacologic properties. Also, ginseng's activities on the CNS are most confusing.[6]

Early pharmacologic studies on ginseng in Japan revealed the rather complex effect of ginseng's properties on the nervous system. For example, *Fujitani* found that panaquilon has hypnotic and depressant activities.[63] *Panax acis* and panacene similarly showed weak sedative and hypnotic effects on the cerebral centers, but they stimulated the vasomotor and respiratory centers, thus increasing blood flow and pressure and the rate of respiration when given in small doses, but lowering the blood pressure when given in large doses,[64-66] Yonekawa[67] investigated the various pharmacologic activities of ginsenin. A stimulant effect on the nervous system in mice was observed using small doses, and a depressant effect was obtained using large doses. Ginseng saponins (extracted from ginseng roots) had stimulatory effects on the CNS in rats, as was shown by decreased sleeping components and increased activity components.[68] Ginseng saponins increased in low doses (1.5–5.0 mg/kg), but decreased in high doses (100 mg/kg), the spontaneous motor activity in mice and rats.[69] Apparently in small doses, ginseng saponins stimulate the CNS, but the mode of action is different from that of an amphetamine or Benzedrine-like compound.[70] The stimulant action of ginseng is characterized by low toxicity, absence of pronounced excitation, and lack of disturbance of normal sleep. Ginseng is a remedy having few side effects in comparison to other known chemical stimulants.[1]

Again, from the studies of ginseng aqueous extract, the Chinese scientists found that ginseng was a potent sedative.[20] It weakens respiratory systems, slows spontaneous activity, and also prevents convulsion provoked by a number of CNS stimulants, such as caffeine, sodium benzoate, strychnine nitrate, and Cardiazol (pentylenetetrazol).

Tagaki and his associates at the Faculty of Pharmaceutical Sciences, University of Tokyo, conducted a series of systemic pharmacologic studies on ginseng extracts and ginseng saponins.[71,72] The crude ginseng extract was separated into several fractions, and each fraction contained two to three individual saponins. The first fraction (G No. 3, containing ginsenosides Rb and Rc) showed CNS depressive, ataractic, analgesic, and muscle-relaxant activities in mice, rats, and guinea pigs. The second fraction (G No. 4 containing ginsenosides Rg_1, Rg_2, and Rg_3) had both stimulant and depressant activities, in addition to muscarinic and histaminic actions. The third fraction (G No. 5 containing lipophilic saponins) had papaverine-like in addition to muscarinic and histaminic actions. Fractions No. 4 and No. 5 had central depressant action, such as decreased spontaneous movement, decreased body temperature and alertness, and relaxed muscle tone. Accordingly, ginsenosides of the protopanaxadiol series (Rb group) are sedatives, while ginsenosides of the protopanaxatriol series Rg group are mostly stimulants. This explains why ginseng extract contains both *depressant* and *stimulant* components.[71] Kaku et al. further investigated the pharmacologic properties of each of the individual purified ginseng saponins, Rb_1, Rb_2, Rc_1, Rd, Re, Rf, Rg_1, and Rg_2. Their data show that all the saponins exhibited antifatigue action. They markedly increased the movement after compulsory gait, and the action was consistent and independent of their action on the movement before compulsory gait. These saponins also showed moderate depressant actions on electroencephalogram and in the behavior of cats. Saponins Rg_1, Re, and Rb_2 are more potent than the others.[73]

Many clinical tests in the Soviet Union in the early 1950s revealed that ginseng extracts are effective for patients who are suffering from different types of nervous exhaustion and several emotional disorders.[1,74] After treatment with ginseng, the patients showed weight gain with the disappearance of common symptoms such as weakness, fatigue, headache, insomnia, distress, and uneasiness. Ginseng extract also produced good results in a number of other emotional disorders such as mental weakness and exhaustion, various emotional disorders including vegetative nervous disorders, hypertension, gonadal disorders, hypotension, and loss of appetite. It seems from clinical experience that ginseng extract would be useful as a supplementary remedy for emotional disorders.[73]

IMPROVING MENTAL EFFICIENCY AND WORKING CAPACITY

The influence of ginseng on working capacity and mental efficiency is a new dimension of pharmacologic research. That ginseng increases memory has been claimed by the ancient Chinese doctors. According to Brekhman, however, "The fact that ginseng increases both mental capacity and quantitative indices of work in laboratories and at industrial occupations indicates its powerful effect upon a specific portion of the central nervous system." It was assumed that ginseng may have a particular and, most probably, electrical effect on the highest sections of the CNS.[6] Ginseng has a favorable effect on the cerebral cortex process and makes the conditioned reflexes easier; thus, ginseng is capable of increasing the mental efficiency of man.

Many studies have been conducted in the Soviet Union and in Sweden to prove that ginseng increases mental efficiency. In proofreading, all the tests showed that ginseng helped to lessen the number of mistakes (by about 51%), while the number of letters read was only slightly increased (by about 12%). Ginseng extract was also tested in telegraph operators who had to transcribe a special code in a fixed period of time. Ordinarily, the continued work would result in an increase in the number of errors of about 128%, while in the operators who took 30 mL of ginseng extract, the number of mistakes decreased by about 38%.[6]

In a recent study in Stockholm, healthy college students served as volunteers in a double-blind experiment lasting for 33 days. The volunteers were divided into three groups, each group containing five male and five female students. The students were treated with two ginseng capsules per day. The tests conducted were the spiral maze test and the letter cancellation test. At the end of the 33-day tests, the students given ginseng extract (Geriatric-Pharmaton or Gerikomplex-Vitamex)[75] treatment had fewer errors than the control group who were treated with placebo capsules. Also it was found that the two ginseng preparations exerted an almost equally favorable effect on psychomotor function and on mental capacity. In other words, students given ginseng showed a statistically significant increase in mental concentration.[76]

Accordingly, a healthy individual may increase his capacity for intellectual work in addition to physical performance under the influence of ginseng. The improvement of work is qualitative but not quantitative; in this regard, ginseng may increase industrial productivity. This is, as Brekhman called it, "the tonic effect on man's organism" to prove that ginseng is a remedy that when taken for some time, causes increased vitality, leading to greater work efficiency and mental capacity.

Recent clinical studies on mental performance of ginseng by Hallstrom, Popov, Lo, and D'Angelo showed that

- Ginseng improved competence, mood, and performance of nurses on night duty.
- Bulgarian research showed improvement in mental and nervous problems without adverse side effects.
- In aged patients with problems of reduced intellectual output, loss of memory, and slow thinking, the therapeutic effect of ginseng was seen after 1 week with considerable improvement noted after 2 months.
- In a double-blind crossover trial on university students, ginseng improved performance of mental arithmetic.

REGULATING YOUR BLOOD PRESSURE

Hypertension is one of the most common human disorders and a major cause of complications leading to death in middle-aged and aged individuals. At least 23 million people in the United States have hypertension, which is a major risk factor for strokes and heart failure. There is a substantial increase in hypertensive heart disease with age in both sexes. Blacks have substantially more heart disease and hypertension than whites, and usually males have more than females.

How Ginseng Acts on the Cardiovascular System

Although it is premature to say that ginseng is effective in its treatment or prevention of hypertensive or hypotensive heart diseases, the effects of ginseng on the heart, blood pressure, blood vessels, and even respiratory systems are medically significant. Quite a number of studies have been reported in medical literature. The sedative effect of ginseng, that is, lowering the blood pressure, has been the most interesting and pharmacologically useful observation.

Yonekawa[66] observed that ginsenin increased the function of an isolated frog's heart at a small dose, but paralyzed it at a large dose. Kin[51] found similar effects with panax acid. In studies with nonisolated hearts of frogs and cats, ginseng showed similar dose-dependent stimulant-depressant effects. In other studies, ginsenin,[66] panaquilon,[62] and panax acid[64,65] all showed blood pressure lowering (depressor) but not pressure increasing effects in animals. Kin[51] and Watanabe,[77] on the other hand, reported that ginseng and ginseng saponin caused an increase in blood pressure (pressure effect) at small doses, but a pressure lowering (depressor effect), which remained for some time, occurs at large doses in rabbits. Injection of ginseng extract also caused a depressor effect in anesthetized dogs.[78]

Toward the blood vessels, the alcoholic extract of ginseng initially contracts, then dilates, the capillary blood vessels in frogs. Ginsenin showed a vasocontraction effect in frogs and rabbits at small doses, but a vasodilation effect was observed at large doses.[66]

Administration of ginseng saponins, ginsenosides, to dogs caused a remarkable increase in blood flow, and the effect was more potent than that induced by papaverine. Ginsenosides also showed a vasodilator effect. The ginsenoside Rg_1 at a dose of 5–10 mg/kg had a depressor effect preceded by a slight pressor effect, but the respiration was little affected. The dose-dependent raising-lowering of blood pressure (biphasic effect) seemed to be due to the direct action of the ginseng saponin on the blood vessels.[72] Because of the vasodilation effect of the capillary vessels, ginseng extract also increased the blood supply to tissues and the brain.[79]

The depressor effect, as a matter of fact, has been adopted in China as well as in the Soviet Union as an indicator in biological analysis for potency determination of ginseng preparations.[55] In other words, the potency of a ginseng preparation can thus be standardized. The practical application of the depressor effect in humans after an oral dose of ginseng, however, is still questionable, and more research definitely is needed to prove it.

Other studies in China showed that ginseng has direct dynamic action on the blood flow, with an increase in the force of cardiac contraction, especially during acute circulatory failure or shock.[80] The effect of ginseng in normalizing the level of arterial pressure is indeed unique. It is effective in the treatment of abnormal hypotension. Clinical research conducted at Hwa-san Hospital in Shanghai showed interesting results. Seven cases of acute interference with blood flow of the heart muscle (myocardial infarction) with shock irregularity in the rhythm of the heart's beating were involved. The patients were treated by a routine method or with ginseng. When a Western type of drug was given to raise the blood pressure, the blood pressure rose only temporarily, then fell in a short time even after repeated doses. However, when ginseng preparations, either alone or in conjunction with *seng fu tang*, were given to similar patients, the blood pressure rose and remained at a normal level. In addition ginseng also prevented shock, and patients returned to normal.[81]

By contrast, clinical work in Asia is carried out with far higher amounts of ginseng and for uses that differ markedly from those described in the West. The *Pharmacopoeia of the People's Republic of China*[82] officially lists 3–9 g as the dosage for ginseng in decoction (tea) form. Thus, for example, in the attempt to prevent and treat cardiovascular diseases, ginseng is given in this dosage range to lower blood pressure. By contrast, Western literature cautions persons with hypertension to avoid ginseng, especially to avoid red ginseng, even in the much lower doses used in the West. In a recent report from Korea,[83] red ginseng was administered at a dose of 4.5 g per day (Korean red ginseng powder, 1.5 g per dose, three times daily) and reported to have a slight blood pressure lowering effect (about 5% average decline) after 2 months daily administration. In reviewing prior studies of ginseng's effect on hypertension, the authors of that study noted that there had been negligible or minor effects on blood pressure previously mentioned for administration of 3 g red ginseng powder every day for 3 months and with 3–6 g of red ginseng powder every day for an average duration of 10 months.

As a reflection of this direction in ginseng research, in 1980 the Institute for Traditional Medicine of Portland, Oregon, conducted a clinical trial in Santa Cruz, California, on the impact of Asian ginseng on the risk factors for cardiovascular disease: cholesterol, triglycerides, elevated blood sugar, and blood pressure.[84] In that study, the first clinical trial of ginseng in the United States, 100 patients received red ginseng provided by the Korean Ginseng Research Institute at a dosage of either 3 or 4.5 g per day (others received a placebo). Only modest effects were observed after 3 weeks of daily administration of the ginseng capsules, with slight favorable improvements in the risk factor.

GINSENG CONTROLS BODY'S CHOLESTEROL AND ATHEROSCLEROSIS

Cholesterol is produced naturally in our body and exists in blood, bile, and the gallbladder. Another direct source of cholesterol is foods such as beef drippings, milk, eggs, cheese, butter, cream, organ meats, etc., all of which contain high levels of cholesterol.

The cholesterol content in blood is usually maintained at a constant level. For example, if dietary sources are high in cholesterol, less cholesterol will be formed in the liver. A number of hormones markedly affect the formation of cholesterol and its degradation. For example, insulin and thyroid hormone deficient people usually suffer from high cholesterol.

According to Dawber and Kannel of the Framingham Heart study, a man under 59 years with a blood cholesterol measurement of 260 or above has more than three times the risk of heart attack than a man with a blood cholesterol level below 200.

Laboratory testing animals—monkeys, dogs, cats, pigs, chickens, rats, guinea pigs, and hamsters—fed a high-fat and high-cholesterol diet, developed atherosclerosis and heart attacks, resulting in deaths sooner, and in more instances, than in those fed with a low-cholesterol diet. This is similar to what is observed in humans.

At present, atherosclerosis, particularly of the coronary arteries, is the most widespread illness in the United States. In 1965, nearly 40% of all deaths in the United States were due to heart attacks caused by atherosclerosis. Of these deaths, nearly one-third occurred in people under the age of 65. Today, about one in five American males will develop coronary disease before the age of 60 years. The chance that he will die of the heart disease before 60 years is about 1 in 15.[85]

If you have high blood cholesterol, your doctor may have told you to eat a low-cholesterol and low-fat diet. You may have known that in order to reduce cholesterol in the diet you have to use unsaturated vegetable oils for cooking, instead of saturated oil, animal fat, butter, or lard; switch from deep-fried foods to boiled foods; and eat less cheese, milk, egg yolk, liver, and butter. You may also want to switch from regular cheese to cottage cheese, whole milk to skim milk or even soybean milk, and eat more fish, chicken, and turkey instead of beef, pork, duck, and bacon. You may even want to cut your regular American diet with Chinese dishes, which contain less cholesterol and less calories.

GINSENG CONTROLS CHOLESTEROL LEVEL

From animal studies, ginseng saponins show a marked effect on the regulation of blood cholesterol levels. After administration of ginseng, animals showed an increase in the synthesis of cholesterol in the liver, but the total cholesterol concentration in the serum decreased by about 25%, as a result of decreased absorption of cholesterol from the gastrointestinal tract and also an increase in the speed of cholesterol metabolism in the animal's body. In addition, abnormal (high) cholesterol levels in the blood and deposits of fat in the heart can be prevented when ginseng is included in the diet.[86]

The effect of ginseng on blood cholesterol levels in connection with diet and antisclerotic drugs was investigated. Popov of the Revitalization Center at Nassau, Bahamas, found that with ginseng preparations in combination with antisclerotic drugs, the cholesterol levels of the blood fell nearly 20% in the treatment of 106 patients aged from 40 to 70 years, suffering from hypertension and high blood cholesterol. Popov concluded that ginseng extract possesses a substantial synergistic effect in reducing blood cholesterol concentration.[87]

EFFECT ON LIPID METABOLISM

Rates receiving an intraperitoneal dose of 50 mg/kg of ginsenoside Rc showed a threefold increase of epididymis fat synthesis from ^{14}C-acetic acid relative to the control, and a sixfold increase in intestinal fat synthesis. The lipolytic effect of adrenocorticotropic hormone and insulin could be antagonized by most ginsenosides, including ginsenosides Rb_1, Rb_2, Rc, Rd, Re, Rg_1, Rg_2, and Rh_1.[88]

The cholesterol and triglycerides levels in plasma in rats fed a high-cholesterol diet were significantly decreased by intramuscular administration of ginsenosides at 10 mg/kg for 14 days. But the biosynthesis of cholesterol was increased, and the turnover of cholesterol into bile acids and the excretion of cholesterol were also promoted to keep the lower level of cholesterol in the blood. It was also found in rabbits fed a high-cholesterol diet that ginsenosides promoted the metabolism of cholesterol and lowered blood cholesterol levels to prevent aortic atheroma formation. Ginsenoside Rb_2 remarkably decreased the total cholesterol and low-density lipoprotein (LDL) cholesterol, and significantly increased high-density lipoprotein (HDL) cholesterol in the blood of rats fed a high-cholesterol (1%) diet. In an experiment using ^{14}C-acetic acid incorporation, ginsenoside Rb_1 was found to significantly promote the biosynthesis of cholesterol in the liver and serum of rats.[89]

ANTIDIABETIC EFFECT OF GINSENG

Diabetes is a universal disease that is encountered most frequently in industrialized countries, particularly among poor and older people. Despite the important medical advances in the treatment of the disease and its complications, diabetes has accounted for about 38,000 deaths a year in the United States since 1965, and has been implicated as a contributory cause in a substantial proportion of deaths from other chronic diseases. Diabetes has been characterized by consistently higher mortality rate among females than among males.[90] It is more interesting to note that, from 1964 to 1973, the mortality rate from diabetes among men insured under standard, ordinary policies of the Metropolitan Life Insurance Company decreased by 20%, while that among insured women increased by 5%. By 1999, 450,000 deaths were attributed to diabetes, which has now become the nation's third-ranking cause of death, behind cardiovascular disease and cancer, and is increasing rapidly at a rate of about one million new diabetics every 3 years. The disease is about three times as common among the poor as among middle-income families and the rich.

In diabetes the metabolism of carbohydrates is impaired, and that of proteins and fats is enhanced. Glucose in the body thus builds up to high levels, especially in the blood (hyperglycemia), and is excreted in the urine (glucosuria). The breakdown of tissue proteins, lipids, and fatty acids is sharply accelerated in diabetics, thus producing a large amount of nitrogen and ketone bodies in the urine. The excretion of glucose and ketone bodies causes a big loss of body water and salt, thus producing a dehydration effect and a feeling of great thirst. In severe cases, it can produce ketoacidosis (a disease of high ketone in the urine), coma, and eventually death, if the condition is not corrected by the administration of insulin.

Obesity is the most common contributing factor to diabetes. Heredity also appears to be an important factor. The life expectancy of diabetics is estimated to be shorter by one-third than that of nondiabetics. Failure to control the diabetic state may lead to serious complications. Cataracts and blindness, leg gangrene, heart disorders, including atherosclerosis, and kidney failure are the most common diabetic complications. Diabetes is also the second leading cause of blindness in the United States.

Drugs lowering the blood sugar level are called *hypoglycemics*. Insulin had been the most effective antidiabetic drug. However, it is ineffective orally. Unfortunately most of the clinically available oral hypoglycemic drugs have undesirable side effects. Research findings show that treatment of diabetes with oral hypoglycemic drugs may be associated with an increased rate of cardiovascular mortality, when compared to the treatment of diabetes with diet alone or diet plus insulin.[91] The effect of ginseng on hyperglycemia has been extensively investigated both at laboratory and clinical levels. Preliminary data are promising and interesting.

GINSENG FOR DIABETES

One of the most important medicinal uses of ginseng is its activity for a diabetic patient. Ginseng's activity toward diabetes seemed to be known to Chinese doctors as early as the first century. T'ao Hung-ching's *Ming-I Pieh-lu*, a famous Chinese medical book published before the Han dynasty, described the usefulness of ginseng in eliminating tiredness, thirstiness, and polyurea, which are the most common symptoms of diabetes. Other famous Chinese medical compendia, such as *Hai-yao Pen-ts'ao* written by Li Heng of T'ang dynasty, *Yung-Yao-Fa Hsiang-lun* by Li Tung-yuan of Yuan dynasty, and *Pen-ts'ao Kang-mu*, edited by the most famous pharmacologist and physician, Li Shih-chen of the Meng dynasty, recorded the uses of ginseng to relieve the ailment or symptoms related to diabetes.

There are at least 25 separate modern scientific studies of ginseng's activities in diabetes conducted in Japan, China, Bulgaria, the Soviet Union, and Korea. The pioneer Japanese workers in this area, Arima and Miyazaki,[92] who treated high blood sugar (hyperglycemia) in rabbits with ginseng saponin, observed, as with ginseng extract, a blood sugar lowering (hypoglycemic) effect. Kin[93] reported that subcutaneous injection of ginseng saponin, panax acid (an active component of ginseng), or oral administration of panacene (another active component of ginseng) to rabbits produced a marked hypoglycemic effect. Particularly, ginseng saponin inhibited adrenaline-induced hyperglycemia, but questionable results were obtained on hyperglycemia induced by ammonium chloride.

In China during the 1950s C. K. Wang and H. P. Lei of Peking University,[94,95] C. Y. Sung and T. H. Chen[96] of the Chinese Academy of Medical Sciences in Peking, and C. Tsuao and C. C. Yen[97,98] made separate studies on the effect of ginseng in alloxan diabetes. Their data, as were reported to the First Conference of the Society of Chinese Physiological Science in 1956, indicated that the application of ginseng extract decreased blood sugar and urine acetate in alloxan-induced hyperglycemia in dogs, rats, and mice. Later in 1957, Wang and Lei,[94] by using a ginseng preparation (a 50% alcohol extract of red ginseng and dried) to normal and alloxan-induced diabetic male dogs. With daily doses of 0.13, 0.2, and 1.0 mg/kg of body weight, for 2 weeks, observed the hypoglycemic effect of ginseng when given alone or in combination with protamine-zinc insulin. The average blood sugar levels in the period 5–10 days prior to ginseng administration were compared with that during the last day of treatment. In the normal dogs, the administration of 1 mg/kg of ginseng preparation produced somewhat drops in the blood sugar levels. A dosage of 0.2 mg/kg produced no significant change, but in another dog it produced a rise in blood sugar level. In the diabetic dogs, the ginseng preparation at 1 mg/kg produced a significant decrease in blood sugar levels. Apparent falls in the urine output and urine sugar were sometimes observed during the treatment. Although the dose level of 0.2 mg/kg failed to change the blood sugar level, in every instance, the general condition and urine ketone reactions were improved during the period of ginseng treatment. In spite of the fact that ginseng had a hypoglycemic effect (at high doses) in the normal and diabetic animals, it was unable to completely correct the metabolic disturbance occurring in alloxan-diabetic dogs and, therefore, failed to check the diabetic state. It may be helpful in the treatment of the diabetic state, but it cannot be relied upon to act as a substitute for insulin.[55]

In 1959, Tsuao, Yen, and Lei further reported that ginseng normalized blood sugar levels in alloxan-induced diabetic mice but did not prevent their death.[98] Liu, Chi, and Sung[99] studied the effect of ginseng on alloxan diabetes in rats fed different diets. In the first series of experiments, 17 rats were divided into two groups and were given alloxan, 200 mg/kg, intraperitoneally. One group of the animals (6 rats) was fed with cornmeal, a high-carbohydrate diet, plus 5% of ginseng powder in the cornmeal, while the second group (11 rats) was fed only with cornmeal for a total of 12 weeks (1 week prior to and 11 weeks after the injection of alloxan). After 3 days of alloxan injection, 3 of the 11 rats in the control group were dead, while only one rat in the ginseng-fed group was dead. Of the remaining eight rats in the control group, six rats developed diabetes, the blood sugar levels were about 557 ± 17 mg%, and 7 weeks later, all of the six diabetic rats were dead. In the ginseng-fed group, however, among the six rats, one was dead and three developed mild diabetes; the average blood sugar levels were 587 ± 75 mg%. The rats continued with 5% ginseng powder in the feed gradually lowered their blood sugar levels to normal and remained at normal blood sugar levels (ca. 150 mg%) for at least 3 weeks after the withdrawal of ginseng treatment. As shown in Table 16.3, the alloxan-induced diabetic rats' high blood sugar levels reduced to normal 7 weeks after the continued administration of 5% ginseng in the diet, while the diabetic rats of the control group were all dead within 7 weeks.

In another series of experiments wherein the rats were fed a high-protein diet (a mixture of fish powder, beans, whole wheat, bone meal, yeast powder, and cod

TABLE 16.3

The Effect of *Panax ginseng* on Alloxan-Diabetes in Rats Fed with Cornmeal

	Week											
	Rats	1	2	3	4	5	6	7	8	9	10	11
5% ginseng	1	339	510	176	146	116	130	168	112	125	140	83
powder	2	600	600	589	600	584	364	160	128	245	214	190
added in	3	463	420	294	255	208	274	160	170	146	152	148
diet	4	197	174	128	139	148	160	144	141	175	130	102
	5	170	144	143	139	130	136	152	130	136	116	128
Control	1	554	Dead									
(no ginseng)	2	585	Dead									
	3	524	600	Dead								
	4	600	600	Dead								
	5	600	600	600	600	600	Dead					
	6	600	600	576	600	600	600	Dead				
	7	167	162	122	126	116	144	120	130	110	80	101
	8	222	198	145	143	152	176	162	130	140	124	101

Column header: Mean Serum Sugar Levels (mg%)

Source: Data after Liu, K. T et al. *Acta Pharm. Sinica*, 7: 213, 1959.

liver oil), the blood sugar levels of the alloxan-diabetic rats treated with ginseng did not differ significantly from those of the untreated diabetic rats. It appears that the *diet* of the animals made the results of the two series of experiments contradictory. Nevertheless, when the high-protein diet was replaced by cornmeal during the latter half of the experiment, the blood sugar levels of the untreated rats increased drastically while no significant change in blood sugar levels was observed in the ginseng-treated rats. This experiment clearly indicated that although the rats fed with 5% ginseng for 1 week could not escape the action of alloxan, ginseng, ultimately, did normalize the blood sugar levels of alloxan-induced diabetic rats, particularly when the rats were fed a high-carbohydrate diet.

Studies in the Soviet Union and Bulgaria in the 1950s and 1960s further supported the theme of ginseng's uses in diabetes. In the treatment of alloxan-induced diabetes, ginseng was found to impede a loss of body weight of rats, to reduce the sugar levels in the urine, and to prolong the survival time of the diabetic animals.[100] Another study showed an extract of ginseng reduced the high blood sugar levels in rabbits with adrenaline hyperglycemia, and in humans with alimentary hyperglycemia.[101]

According to Petkov,[102] ginseng evidently is capable of potentiating the action of insulin. It is very surprising that ginseng extract regulates the carbohydrate metabolism both ways—that is, in an experimentally induced hyperglycemic state, ginseng was able to reduce the blood sugar level, while in an insulin-induced hypoglycemic state, ginseng could bring about an increase in the blood sugar level to normal.[103]

From biochemical approaches, the effect of ginseng on carbohydrate metabolism has been evaluated by many investigators. Lee[104] found that ginseng may contain a substance that is capable of increasing the basal metabolism in normal rats; in this regard, the effects could be blocked by antihistaminic agents. Based on this finding, it was believed that the effect of ginseng on metabolism may be due to histamine liberation, or that ginseng may stimulate the liberation of histamine in the body.

In 1973 Oura and Hiai[105] of Toyama University reported that after intraperitoneal injection of ginseng extract (a fraction 4), the liver glycogen and the reducing sugar levels in rats were reduced. Ginseng saponin contained in the extract somehow caused a decrease in the total carbohydrate level in the animal. Ginseng extract caused a marked reduction in the blood glucose levels of rats with adrenal glands removed, but had a weak or insignificant effect in normally fed rats and infested rats infused with glucose solution. The reduction of blood sugar was observed 2 hours after the injection of ginseng, and this effect continued for about 8 hours. Also, in rats, liver glycogen content and the sugar levels in the liver, kidney, and muscles were also remarkably reduced after the treatment with ginseng extract.[106] The authors suggested that ginseng extract turns the metabolic flow in the direction of lipogenesis by the conversion of sugar. This was confirmed by the fact that the first phenomenon observed was a striking stimulation in the incorporation of radioactivity into the adipose tissue beginning 2 hours after the administration of ginseng. Almost simultaneously a slight decrease of blood sugar level was also observed.[107] The investigators also found that the glycogen content of rat liver varies with diet. Similarly, from the application of a radioisotope, sodium ^{14}C-acetate, the incorporation of the ^{14}C-acetate into the total lipid and glycogen content in the liver was determined 4 hours after injection of ginseng extract into rats The results indicated that a striking stimulation of metabolism was obtained in rats fed a fat-free diet over the high-fat diet group.[108]

The early clinical studies on the hypoglycemic effect of ginseng were conducted in Korea and Japan in the 1930s. In the treatment of diabetic patients, Saito[109] and Saito and Abe[110] found that the alcoholic extract of ginseng was effective in inhibiting adrenaline hyperglycemia and alimentary hyperglycemia. In Japanese hospitals in the early days, diabetic patients were treated with ginseng powder or by injection with ginseng extract. The blood sugar levels and the general symptoms of the diabetic patients were indeed reduced, but a cure of the diabetes was not obtained.

In the Soviet Union, Shass[111] reported that the treatment of diabetic patients with 70% alcoholic extract of ginseng gave good results, and Shass even suggested that ginseng could substitute for insulin. The hypoglycemic effect was very good during winter and fall seasons, but not so good during the spring and summer seasons.

Recent clinical studies in China showed that ginseng was capable of lowering the blood sugar levels to about 40–50 mg% in diabetic patients. This hypoglycemic effect continued for about 2 weeks after the treatment. Unfortunately, ginseng, like other chemical hypoglycemic agents on the market, exerted no curative effect in moderate and severe diabetic patients; general symptoms such as thirst, weakness, and polyurea, however, were much improved, or disappeared. Ginseng cannot substitute for insulin in patients who use insulin for treatment, the insulin dose level could be decreased if ginseng is concomitantly administered.[55]

During the period of January to July of 1970, T. Kikutani of Sanraku Hospital, Tokyo, investigated the effect of ginseng on diabetic patients clinically. Twenty-one diabetic outpatients (13 male and 8 female) were the subjects, and they all complained of having common subjective symptoms of fatigue and malaise, hypertension, thirst, proteinuria, increased intake of water, loss of energy, coldness, and lumbago. These patients were treated with a ginseng preparation in capsules (a commercial product of Yamanouchi Drug Co., Japan), one capsule each time, three capsules a day for 4–6 months. Each capsule contained 100 mg of dried, powdered ginseng extract and 20 mg of vitamin E. The blood sugar levels and other symptoms of the patients were examined periodically. From this pilot clinical study, it was found that the ginseng extract was effective on 15 out of the 18 diabetic patients. Their subjective symptoms were improved and their blood sugar levels were reduced to normal along with the treatment.[112]

In another clinical trial, Kakiochi, Imanaka, Kohuo, Ohara, and Nakajima of Osaka University, School of Medicine, treated 18 diabetic patients with the same ginseng preparation as that used in the studies previously described. However, in this study, the patients were allowed to take two capsules each time, three times a day—that is, each patient takes 600 mg of ginseng extract and 120 mg of vitamin E per day. The duration of the treatment varied from 14 to 84 days depending on the conditions of the patient. The patients were questioned about their symptoms regularly, and their blood, stool, and urine samples were analyzed, and their renal and hepatic functions were checked by the staffs of the same hospital. The results showed that ginseng was effective in four out of seven patients against general *malaise*, three out of five patients against tiredness and fatigue, four out of five patients against impotence, and three out of three patients against thirst. In other words, the ginseng extract at 600 mg level daily dose was about 60%–100% effective in the treatment of the subjective symptoms of the diabetic. Also found was that ginseng was able to improve the anemic conditions of the patient. In insulin-resistant patients, ginseng was able to reduce the dose level of insulin if ginseng was concomitantly used. The investigators also indicated that the 600 mg dose of ginseng may be required in the onset of the treatment, but a 300 mg maintenance dose may be sufficient for chronic patients who need a long-term therapy. In both of the studies, no particular reason was given why vitamin E was added in the ginseng capsules.[112]

REMARKABLE HEMATINIC EFFECT OF GINSENG

ANEMIA

When the number of red blood cells or the amount of hemoglobin within them falls below certain levels, a person is said to have anemia. Anemia often causes pallor, weakness, lack of vigor, and most important of all, susceptibility to other diseases. Anemia patients have high mortality rates and, therefore, every effort should be taken to reduce or prevent its occurrence.

The loss of a large amount of blood from the body (usually from a bleeding wound, bleeding from the gastrointestinal tract, an ulcer, or due to abnormal menstruation in women) may cause anemia. A drop in red blood cell production as a result of iron deficiency in diet, malnutrition, organic diseases, or inability to absorb the blood-making elements such as vitamin B12, folic acid, and iron may cause anemia. Degeneration, defectiveness, or short life or red blood cells may also induce anemia. Nevertheless, iron deficiency is the most frequent cause of anemia.[113,114]

GINSENG'S HEMATINIC EFFECT

From animal studies, ginseng acts as a powerful hematinic agent. It increases hemoglobin levels and restores red blood cells quickly in animals after bleeding. Kin[115] observed that rabbits receiving 0.5 g/kg daily of ginseng extract showed a rapid increase in the number of red blood cells, which continued for about a month and then ceased. The hemoglobin content and the white blood cell (leukocyte) counts were also greatly increased. A hematinic iron preparation, Blutone (a strong antianemia agent), given at 13 g/kg to rabbits as a control, induced a lesser increase in red blood cells and hemoglobin than did ginseng extract. Ginseng was thus called a hematinic.

Several other studies showed ginseng causing an increase in the number of red blood cells and hemoglobin in animals.[116] Ginseng extract also showed fast restoration of blood albumin in animals after massive bleeding.[117]

Recently, a group of Japanese investigators, Oura et al., reported that intraperitoneal injection of ginseng extract to rats and mice increased the synthesis of serum proteins.[118–120] Serum proteins synthesis increased by about 46%–49%. Ginseng extract also stimulated the synthesis of ribonucleic acid (RNA) in rat's liver, in addition to the synthesis of bone marrow deoxyribonucleic acid (DNA), and the normal division of bone marrow cells. The active principles of ginseng extract, particularly the ginsenosides, are expected to have a metabolic stimulating or hormone-like action. Biochemical studies have added further positive evidence supporting the thesis that ginseng is a hematinic agent. Regarding the hematinic effect, one of the active principles of ginseng, *prostisol*, plays an essential role in protein and RNA synthesis in the rats' liver.[121,122]

The hematinic effect of ginseng was also demonstrated in humans, particularly those who have been ill for a long time, who are suffering from wasteful or consumptive diseases, or are recovering from major surgery. Following surgical operations for gastric ulcer, patients in a Soviet hospital were given the conventional treatment and care, and a special group was given additional ginseng preparations. It was found that the patients treated with ginseng were the first restored to health and returned to work.[19] In a recent clinical trial at National Shirahama Hot Spring Hospital in Japan, the efficacy of an aqueous extract of ginseng (containing prostisol) on fatigue and anemia was critically evaluated. The ginseng extract was particularly effective in the treatment of anemia due to tuberculosis, old age, and iron deficiency, but ineffective toward pernicious anemia. In addition, the symptoms associated with anemia, such as rheumatoid arthritis, fatigue, and debility, were eliminated. Appetite, hemoglobin level, and the number of red blood cells and platelets were greatly improved or increased. Among the 50 patients treated, no one complained of any side effects with ginseng.[123]

EFFECT ON RNA, DNA, AND PROTEIN METABOLISM

Ginseng extract increased the biosynthesis of RNA and incorporation of amino acids in the nuclei of hepatic and renal cells, resulting in an increase of serum protein. Ginsenoside Rd exhibited the greatest effect on protein synthesis. Experiments also showed that ginsenosides increased the synthesis of high-density lipoprotein (HDL) and apoprotein in serum. The activity of DNA-dependent RNA polymerase in hepatic cellular nuclei of rats increased markedly following administration of ginseng extract. The rate of RNA synthesis in the nuclei and cytoplasm was subsequently increased.

Administration of ginseng also greatly increased the biosynthesis of DNA and protein in the bone marrow. It is thus believed that ginseng is beneficial in the treatment of hemorrhagic anemia.[88]

IMPROVEMENT OR CURE FOR INDIGESTION

When we say indigestion or digestive disorders, it means the digestive system may not be in normal working order. When digestion is so below par that the food remains undigested, it causes the growth and multiplication of putrefactive bacteria—germs capable of decomposing foods—resulting in the generation of foul gases. The most common indigestion disturbance, particularly in the middle-aged and the aged, is the swelling and sometimes painful gases in the stomach and intestinal tract. Serious and prolonged indigestion could lead to malnutrition and other complications, such as weakness, anemia, debility, and consumption.

Ginseng has a stimulant effect on the smooth muscles of the stomach and intestines; perhaps this is why ginseng has long been recognized as a *bitter tonic, stomachic*, and stimulant to the entire digestive system. In animal studies, it was found that ginseng extract caused acceleration of the gastrointestinal movement[124,125] or increased gastric acidity.[126] The extractives of ginseng leaves similarly showed a smooth muscle stimulant effect. Harding, a pioneer American ginseng user and cultivator, once said that "Ginseng combined with the juice of a good ripe pineapple is par excellent as a treatment for indigestion. It stimulates the healthy secretion of pepsin, thereby insuring good digestion without incurring the habit of taking pepsin or after-dinner pills to relieve the fullness and distress so common to the American people."[127]

In traditional Chinese medicine, famous prescriptions containing ginseng, such as *Si Jaen Zi Tang, Bu Zhong Yi Qi*, and *Gui Pi Tang*, are commonly used to treat or regulate digestion problems.

GINSENG BENEFITS WEIGHT LOSS

Ginseng does many things to help human health, but many people do not realize that ginseng is good for weight loss. It gives an energy boost and can help fight fatigue. Because of this, it allows for the person to be more active. One of the big things that it does in the weight loss battle is it regulates blood sugar. It does this by reducing the amount of carbohydrate that is converted into fat. Also it is important to point out that ginseng alone will not help a person lose weight if the person engages in poor dietary choices. It is vital that a well-balanced diet is followed and that the person is active. Otherwise, like the old English proverb says, "don't dig your grave with your own knife and fork."

One more note, we do not recommend that you take dieting to excesses. The danger of bulimia and anorexia is very real for many people, men and women, in today's diet-crazed world that values looking thin. Looking thin and being healthy is wonderful; however, everything needs to be done in moderation. Think of the health benefits of being a normal weight, do not focus on your body's physical appearance. Most of all, be healthy, and the rest will fall into place.

Hopefully, you will see that ginseng can be a key cog in helping weight loss happen naturally. Just do not get wrapped up in it. Be happy and healthy and remember the funny line by humor columnist Dave Barry, when he said "The leading cause of death among fashion models is falling through street grates." So true, Barry. So true.[128]

WOMEN'S HEALTH—WHAT ARE THE GINSENG BENEFITS FOR WOMEN?

Ginseng, in medical herbalism, is an herb commonly used to treat such women's health issues as abnormal menstruation and strong premenstrual symptoms.

Women in perimenopause often look to solutions for producing more estrogen with ginseng. It is, quite famously, thought that this stimulation of estrogen in the female ovaries causes a boost in libido in women. Of course, women, like men, often like it because it can be useful in weight loss.[128]

BENEFITS OF GINSENG FOR HAIR GROWTH

A recent report indicating that ginseng benefits hair growth is an exciting result of research being conducted in Osaka, Japan.

The study suggested that the Ginsenoside Ro carbohydrate was shown to help fight androgenetic alopecia, commonly known as balding, in men. It is thought that this could have widespread cosmetic implications and lead to future drug development for the prevention and regrowth of hair in men with natural herbal ginseng remedies.

BENEFITS OF GINSENG ON SKIN

Drinking ginseng tea is good for refining and rehydrating the skin effectively. It is also thought to promote skin cells regeneration by increasing oxygenation to skin cells. It also has the effect of boosting blood circulation as well as detoxifying the blood, which are very important to human skin health. Free of toxins, the skin can gain a better complexion and therefore a better look.

REFERENCES

1. Brekhman, I. I. and I. V. Dardymov, *Ann. Rev. Pharmacol.*, 9: 419, 1969.
2. Brekhman, I. I., *Medgiz* (Leningrad), 182, 1957.
3. Brekhman, I. I. and I. V. Dardymov, *Lloyda*, 32: 46, 1969.
4. Brekhman, I. I., "Material for the Study of *Panax ginseng* and Other Medicinal Plants for the Far East," *Primorskoe. Knizhnoe Izdatelstvo*, Vladivostok, 5: 219, 1963.
5. Kitagawa, H. and R. Iwak, *Fol. Phamacol. Jap.*, 59: 348, 1963.
6. Brekhman, I. I., *Med. Sci. Serv.*, 4: 17, 1967.
7. Brekhman, I. I., M. A. Grinevich, and G. I. Glazunov, *Soobshch, Dalnevost, Filiala Akad.* Nauk, SSSR, Vladivostok, 19: 135, 1963.
8. Brekhman, I. I., I. V. Dardymov, and Y. I. Dobrjakov, *Farmakol.I Toksikol.*, 29: 167, 1966.

9. Sterner, W. and A. M. Kirchdorfer, *Z. Gerontol.*, 3: 307, 1970.
10. Rueckert, K. H., in *Proceedings of International Ginseng Symposium, September, 1974, Seoul*, Office of Monopoly, Seoul, pp. 59–64.
11. Rueckert, K. H., in *Proceedings of International Symposium of Gerontology*, April 9–12, 1975, Lugano, Switzerland.
12. Takagi, K., H. Saito, and H. Nabata, *Japan J. Pharmacol.*, 22: 245, 1972.
13. Saito, H., Y. Yoshida, and K. Takagi, *Japan J. Pharmacol.*, 24: 119, 1974.
14. Slepmyan, L. I., *Tr. Leningrad. Kim. Farm. Inst.*, 26: 236, 1968.
15. Ed. *Brit. Med. J.*, 4: 251, 1971.
16. Sharman, I. M. et al., *Br. J. Nutr.*, 26: 265, 1971.
17. Lawrence, J. D. et al., *Am. J. Clin. Nutr.*, 28: 205, 1975.
18. *The Med. Lett.*, 17: 69, August, 1975.
19. Brekhman, I. I., *Ind. J. Publ. Health.*, 9: 148, 1965.
20. Tsung, J. Y., Cheng, and S. Tang, *Sheng Li Hsueh Pao*, 27: 324, 1964.
21. Scicenkov, M. V., *Papers on the Study of Ginseng and Other Medicinal Plants of the Far East, Vladivostok*, 5: 241, 1963.
22. *Cancer Cause and Prevention*, U.S. Department of Health, Education and Welfare, Public Health Service, Washington, DC, 1966, p. 1.
23. *1973 Cancer Facts and Figures*, New York American Cancer Society, 1973, p. 7.
24. *J. Am. Med. Assoc.*, 227: 9, March 4, 1974.
25. Doll, R., C. Muir, and J. Waterhouse, *Cancer Incidence in Five Continents*, U.I.C.C., Geneva, 1970, or *Lancet*, January 17, 1976, p. 131.
26. *The Brit Med. J.*, 3: 329, 1974.
27. Leff, D. N., *Med. World News*, 16: 74, 1975.
28. Moertel, C. G., *J. Am. Med. Assoc.*, 228: 1290, 1974.
29. Lazarev, N. V., *Vopr. Onkol.*, 11: 48, 1965.
30. Stukov, A. N., *Vopr. Onkol*, 13: 94, 1967.
31. Mironova, A. I., *Vopr. Onkol*, 9: 42, 1963
32. Jaremenko, K. V., *Mater*, Izuch Ahen'shonya Drugikh Lek. Sredsto Dal'nego Vostoka, 7: 109, 1966
33. Maljugina, L. L., *Vopr. Onkol.*, 12: 53, 1966
34. Stukov, A. N., *Vopr. Onkol*, 12: 124, 1966
35. Lee, K. D. and R. P. Huemer, *Jap. J. Pharmacol.*, 21: 299, 1971.
36. Kent. S., *Geriatrics*, 30: 140, January, 1975.
37. Kent. S., *Geriatrics*, 30: 164, April 1975.
38. Kent. S., *Geriatrics*, 30: 155, September 1975.
39. Kolodny, R. C. et al., *Diabetes*, 23: 306, 1974.
40. Brody, J. E., *The New York Times*, September 16, 1975.
41. Pfeiffer, E., *Geriatrics*, 28: 172, November 1973.
42. Pfeiffer, E. and G. C. Davis, *J. Am. Geriat. Soc.*, 20: 151, 1972.
43. Pfeiffer, E., *J. Am. Geriatr. Soc.*, 22: 481, 1974.
44. Kent, S., *Geriatrics*, 30: 96, December 1975.
45. Rhodes, P., *Brit. Med. J.*, 3: 93, July 1975.
46. Bowers, M. B., Jr. and M. H. Van Woert, *Med. Aspects Hum. Sexual*, 94 July, 1972.
47. Lawrence, D. M. and G. I. M. Swyer, *Brit. Med. J.*.
48. Eiberhof, *J. Pharm.*, Japan, 39: 469, 1905.
49. Cawa, Y. and I. Saito, *Japan Med. World*, 2: 263, 1922.
50. Min, P. K., *Folia Pharmacol.*, Japan, 11: 238–256, 1951.
51. Kin, K., *J. Chosen, Med. Assoc.*, 5: 21, 1931; *Keijo J. Med.*, 2: 345, 1931.
52. Imamura, T., *Ninjmshih (History of Ginseng)*, Kyoto Shibunkaku, 5: 624–634, 1971.
53. Weber, U. *Suddent Apoth-Ztg.*, 78: 645, 657, and 667, 1938.
54. Kit, S. M., *Farmkol I Tokskol*, 25: 629, 1962.

55. Wang, P. H., *Acta Pharm. Sinica*, 12: 477, 1965.
56. Shibata, K., S. Tadokoro, Y. Kurihara, H. Ogawa, and K. Miyashita, *Kitakanto Med. J.* (Japan), 14: 243, 1964.
57. Yamada, M., *Japan J. Pharmacol.*, 4: 390, 1955.
58. Anguelakova, D. M., D. P. Rovesti, and D. E. Colomobo, *Aromi, Saponi, Cosmet, Aerosol.*, 53: 275, 1971.
59. Suh, C. M., B. H. Kim, and I. S. Chang, *Taehan Saengri Hakhoe Chi.*, 7: 37, 1973.
60. Yamamoto, M., *Taisha*, 10: 581, 1973.
61. Kang, J. W., *Choesin Uihak*, 13: 43, 1970.
62. Lixt, C. R., Jin, L. M., Hy, C., "Regulation on energy metabolism and protection on Panax ginseng polysaccharide," *Am J Chio Med*, 37(b): 1139–1152, 2009.
63. Fujitani, K., *Tokyo Igkukai Zasshi*, 2(3): 43, 1905.
64. Sakai, W., *Tokyo Igaku Dai Zasshi*, 31: 1, 1917.
65. Sakai, W., *Iji Shimbun*, 990: 112, 1918.
66. Sakai, W., *Japan Med. Cit.*, 3: 27, 1918 and 5: 6, 1920.
67. Yonekawa, M., *Keijo J. Med.*, 6: 633 and 785, 1926.
68. Sim, S. J. and J. S. Oh, *Taehan Yakrihak Chapchi*, 9: 9, 1973.
69. Hong, S. A. et al., *Taehan Yakrihak Chapchi*, 10: 73, 1974.
70. Lee, S. W., *Taehan Yakrihak Chapchi*, 85, 1974.
71. Nabata, H., H. Saito, and K. Takagi, *Japan J. Pharmacol.*, 23: 29, 1973.
72. Takagi, K., H. Saito, and M. Tsuchiya, *Japan J. Pharmacol.*, 22: 339, 1972.
73. Kabu, T. et al., *Arzneim. Forsch*, 25: 539, 1975.
74. Chu, S. K., *Yao Hsueh T'ung Pao*, 1: 375, 1953
75. Geriatric-Pharmaton and Gerikomplex-Vitamex are ginseng-vitamin preparations of Pharmaton, Ltd., Lugano, Switzerland.
76. Sandberg, F. in *Proceedings of International Ginseng Symposium, September, 1974, Seoul*, Office of Monopoly, Korea, pp. 65–67.
77. Watanabe, S., *Japan Med. Lit.*, 6: 1, 1921.
78. Raymond-Hamet, C. R., *Acad. Sci.*, 23: 3269, 1962.
79. Zyryanova, T. M., *Stimulyatory Tsent. Nerv. Sist.*, Tomask: 37, 1966, via Chem. Abstr. 66: 114460, 1967.
80. Zhou, J. H., F. Y. Fu, and H. P. Lei, "Proceedings of the Second International Pharmacological Meeting," in *Pharmacology of Oriental Plant*, Vol. 7, K. K. Chen and B. Mukerji, Eds. Pergamon Press, NY, 1965, p. 17.
81. Li, C. P., in *"Chinese Herbal Medicine,"* U.S. Department of Health, Education and Welfare. DHEW Publication No. (NIH) 75–732, Washington, DC, 1974, p. 36.
82. Pharmacopoeia of People's Republic of China (English ed). *American Journal of Chinese Medicine* 26(2):199–209, 1998.
83. Han, K. H. et al., "Effect of Red Ginseng and Blood Pressure in Patient with Essential Hypertension," *Han J. Chinese Med.*
84. Dheerananda, *Panax ginseng: A Clinical Study of Its Effect on Risk Factors of Cardiovascular Disease—Bulletin of the Oriental Healing Arts Institute*, Vol. 1, 1983, pp. 1–13.
85. Stamler, J. in *The Encyclopedia Americana*, Americano Corp., New York, p. 611.
86. Karzel, K., in *Proceedings of International Ginseng Symposium, Seoul, September, 1974*, Office of Monopoly, Korea, pp. 49–55.
87. Popov, I. M. in *Proceedings of Symposium of Gerontology, Lugano*, April, 1975, Pharmaton, Ltd., Lugano, Switzerland.
88. Tanaka, S. S., Shoji, O. H., "Chemistry and Pharmacology of Panax," in *Economic and Medicinal Plants Research*, Vol. 1, H. Wagner, H. Hilino, H. R. Farnsworth, Eds., London Academic Press, 1985, pp. 217–284.

89. Huang, K. C., *The Pharmacology of Chinese Herbs*, CRC Press, Boca Raton, FL, 1993, pp. 21–45.
90. Metropolitan Life Insurance Statistical Bulletin, August, 1975.
91. *The New York Times*, July 4, 1975.
92. Arima, J. and S. Miyazaki, *J. Med. World*, 2: 275, 1922.
93. Kin, K., *J. Chosen. Med. Assoc.*, 22: 221, 1932.
94. Wang, C. K. and H. P. Lei, *Chinese J. Int. Med.*, 5: 861, 1957.
95. Wang, C. K. and H. P. Lei, Abstr. The First Congress of Society of the Chinese Physiological Sciences, 1956, pp. 37–38.
96. Sung, C. Y. and T. H. Chen, The First Congress of Society of the Chinese Physiological Sciences, 35.
97. Tsuao, C. and C. C. Yen, The First Congress of Society of the Chinese Physiological Sciences, 35–37.
98. Tsuao, C., C. C. Yen, and H. P. Lei, *Acta Pharma. Sinica*, 7: 208, 1959.
99. Liu, K. T., H. C. Chi, and C. Y. Sung, *Acta Pharma. Sinica*, 213, 1959.
100. Bezdetko, G. N., T. M. Smolina, and L. d. Shuljateva, 11–12 Izdatlstvo Tomskogo Uniersiteta, Tomsk, 1961.
101. Brekhman, I. I. and T. P. Oleinikova, via *Chem. Abstr.*, 60: 163896b, 1964.
102. Petkov, W., *Arch. Exptl. Pathol. Pharmakol*, 236: 289, 1959.
103. Petkov, W., *Arzneimittel-Forsch.*, 9: 305, 1959.
104. Lee, M. S., *Korean Choong Ang. Med. J.*, 2: 520, 1962.
105. Oura, H. and S. Hiai, *Taisha*, 10: 564, 1973.
106. Yokozawa, T., H. Seno, and H. Oura, *Chem. Pharm. Bull.*, 23: 3095, 1975.
107. Oura, H. and S. Hiai, *Proceedings of International Ginseng Symposium*, Seoul, Korea, September 1974, Office of Monopoly, Seoul, Korea, p. 23.
108. Yokozawa, T. and H. Oura, *Chem. Pharm. Bull.*, 24: 987, 1976.
109. Saito, I., *Japan Med. World*, 1: 699, 1921; and ibid., 2: 149, 1922.
110. Abe, K. and I. Saito, *Japan Med. World*, 2: 263, 1922.
111. Shass, E. Y., *Feldsh I Akush*, 11: 1952.
112. Unpublished Clinical Research Reports on Ginseng from Yamanouchi Pharmaceutical Co. Tokyo, Japan. 1975.
113. Bentler, E., in *The Encyclopedia Americana*, Vol. 1, Americano Corp., New York, p. 831.
114. Harant, Z. and J. V. Goldberger, *J. Am. Geriatr. Soc.*, 23: 127, 1975.
115. Kin, K., *J. Chosen Med. Assoc.*, 21: 647, 873, and 1131, 1931.
116. Kim, H. R., *Ch'oesin Uihak*, 15: 70, 1972.
117. Brekhman, I. I., *Proc. 2nd. Int. Pharmacol. Meet., Prague, August 20–23, 1963*, Pergamon Press, 1965, pp. 97–102.
118. Ohasi, K. and H. Oura, Japanese Patent, 7031, 314, October 9, 1970.
119. Oura, H. and S. Nakashima, *Chem. Pharm. Bull.*, 20: 980, 1972.
120. Oura, H., S. Hiai, and H. Seno, *ibid.*, 19: 1598, 1971.
121. Oura, H., *Japan J. Clinical Med.*, 25: 2849, 1967.
122. Shida, K., *Clin. Endocr.*, 18: 773, 1970.
123. Arich, S., *Taisha*, 10: 596, 1973.
124. Kitagawa, H. and R. Iwaki, *J. Yakurigaku Zasshi (Japan J. Pharmacol.)*, 59: 348, 1963.
125. Petkov, W., *Arzneim-Forsch*, 11: 288 and 418, 1961.
126. Ko, Y. W., *Korean J. Intern. Med.*, 12: 187, 1969.
127. Harding, A. R., in *Ginseng and Other Medicinal Plants*, Emporium Pub., Boston, 1972, p. 167.
128. https://www.sandmountainherbs.com/articles/benefits-of-ginseng.html.
129. Jang, D.-J, M. S. Lee, B-C. Shin, Y-C. Lee, and E. Ernst, *Br. J. Clin. Pharmacol.*, 66: 444–450, 2008.

17 Healthy and Successful Aging

LONGEVITY STORY

A remarkable article entitled "Every Day Is a Gift when You Are over 100" appeared in the January 1973 issue of *National Geographic Magazine*.[1] It represents the oldest people in the world of that time, most of them well over 100 years of age. In the article, Alexander Leaf of Harvard University found that the greatest number of people over 100 years old living in the Caucasus of southern Russia. If we can live healthily, it has been said that the maximum life expectancy of the human species is about 140 years. In other words, you can live an extra 60 or even 80 years if you can live happily and healthily. Would you call it an extra gift from God?

It would be wonderful to preserve a youthful body for as long as possible, but how would you reach this goal? Slowing down aging, of course, is the way to do it. You do not look young if you do not feel young. To feel young means to be healthy. Why are the native Abkhasians of the Caucasus Mountains long-lived? Because they live a way of life that is close to nature. They use all-natural substances, no artificial or synthetic chemicals. Their foods, fruits, and vegetables are organically grown. They use herbal medicine, herbal tea, honey, garlic, and do not eat much meat. They also enjoy physical activity in their advanced years.

The history of the world is full of tales of individuals trying to stave off aging and death. Since antiquity, from ancient Egyptians to Ponce de León, people have explored ways to extend their life spans. In ancient China, it was said that the emperor, Qin-shi huang (221–207 BC), once dispatched 3,000 young men and women to the faraway eastern island of Ponglai to search for a miraculous herb for his dream of eternal life. Unfortunately, their search ended after years of work. Nothing was found.

There have been people who drank gladiator blood or injected themselves with concoctions made from the testes of monkeys. In modern times, people turn to megadoses of vitamins, drink Kombucha tea, use Coenzyme Q_{10}, and inject human growth hormones. Today, Oriental people prefer ginseng root for longevity.

All this is done in the hopes of finding the "fountain of youth" and wishing for a long life. Have these strategies really worked? This chapter is designed to provide information to those who are looking for strategies on how to beat Father Time, explanations for the causes of aging, life cycle, life span, and factors of longevity for enjoying life to its fullest.

DREAM OF ETERNAL YOUTH AND LONGEVITY

We all recognize aging when we see sagging skin, facial wrinkles, gray hair, muscle loss, increasing body fat, hearing loss, poor eyesight, faulty memory, and slower thinking and reactions. These are just a few of the manifestations of aging. While it is easy to point out an aging person, it is not so easy to define what aging is and why one person lives only 40 years while others live to 100 or more.

There have always been stories and legends about miraculously long lives and people who regained their youth. Old legends told of magic potions that kept people young. How do people prevent getting old and live longer? These questions fascinate people, especially the scientists who study aging.

In the second century AD, Egyptian craftsmen in Alexandria were the first alchemists who tried desperately to convert metal into gold. Almost simultaneously and independently, the Taoist monks (not pure alchemists but religious magicians) believed gold to be a miraculous medicine, and they, too, sought to produce it, not for wealth but for perpetual youth and immortality. Some high priests of Taoism dedicated their whole lives to searching for the "life-prolonging elixir."

Poets in the Middle Ages wrote about the Fountains of Youth in which one could bathe and stay youthful. Those ancient alchemists (whose practice was essentially magic) also pursued the dream of eternal youth. They made it one of the three great goals for mankind: to change lead into gold, to travel to the moon, and to discover an elixir of life to keep people young forever.[1]

The ancient Chinese people have a long history of interest in longevity and the search for immortality. Legends exist of *Xian* (an immortal celestial being) who lived forever by mastering longevity techniques. Taoism, the central religious philosophy of China, started as early as sixth century BC. *Lao Tzu*, the old master and spiritual father of Taoism and supposedly the founder of natural philosophy, was only a manifestation of the much older Universalism. The word "Tao," which later became the shibboleth (catchword) of a separate creed, Taoism, is basically a concept common to all Chinese. Taoism is a religio-philosophical tradition that has, along with Confucianism and Buddhism, shaped Chinese life for more than 2,500 years.[2]

Unfortunately, neither the Egyptian alchemists nor the Taoists achieved their dreams. We still cannot turn lead into gold and did not find the magic potion of life, but amazingly, we have reached the moon. As for the dream of staying young and achieving immortality, as time has passed, and modern science and medicine advanced, it is definitely and slowly coming true. Do you not think that those who have lived up to 100 years or beyond are the blessed and fortunate ones in the world?[1]

WHAT IS THE AGING PROCESS AFTER ALL?

Is aging a debilitating disease? If so, aging may be treatable, like any other disease. The truth is that aging is not a debilitating disease. It is a natural, continuous, intrinsic, universal process, and a progressive human phenomenon. No one can stop it, and no one can prevent it.

The aging process is complicated. It involves the effects of heredity (genetic causes), as well as acquired factors, including lifestyle, eating habits, physical activities, health, and diseases. Most likely, the aging process is due to cumulative damage to cells and tissues, which surpasses the body's ability to repair the damage, and to the eventual wearing down of organs and systems of the entire body.

Control of the aging process lies not only in the hands of God but also in the hands of men. Leonard Hayflick, an authority on aging and a Professor of Anatomy at the University of California, San Francisco, said, "Aging is not merely the passage of time. It is the manifestation of biological events that occur over a span of time. Each person's biological clock ticks at a different rate." This means that there is no way to compare *biological age* to *chronological age*. For example, a person who is chronologically 80 years old may be considered to be 60 years old biologically due to sound physical and mental conditions, and excellent health.[3] In contrast, another person who is 60 years old chronologically may be biologically 80 years old, diseased, incapacitated, and suffering. Why can some people run in a marathon at the age of 75 while others are in a wheelchair or handicapped at the early age of 45? The differences involve personal choice of lifestyle and how we live.

Doctors call the normal process of aging "senescense," meaning to grow old or the state of being old. Some people who grow old may have physical or mental difficulties, and doctors call this aging process "senility," to describe problematic aging. The difficulties of senility are most commonly associated with serious impairment of mental functions, such as dementia or Alzheimer disease.

Aging, as a matter of fact, may be classified into two types, namely, *physiological aging* and *pathological aging*. Physiological aging includes all the universal physiological degenerative changes, which develop after an organism reaches maturity (from 20 to 45 years of age). Pathological aging means that as one ages, there are some unavoidable pathological conditions (diseases that develop along with the physiological changes). Pathological changes, no doubt, speed up the normal aging process.[4] The severely polluted environment we are living in today; externally detrimental physical, chemical, or biological factors with which we are afflicted daily; plus our modern sedentary lifestyle, poor nutrition, and stressful life will cause additional pathological changes and speed up aging.

Is there anything that we can do to block the acquired harmful factors of pathological aging? Yes. Adopting a healthy lifestyle, good food habits, disease prevention, and a happy mind are the primary requirements for delaying the aging process. Roger J. Williams, author of *Nutrition Against Disease*, said, "Since aging is inevitable, all we can hope to do is delay the aging process by taking advantage of the understanding of the mechanisms involved. Providing cells and tissues with the best possible nutritional environment would seem to be an obvious expedient."[5] Evidence of the relationship between nutrition and aging is abundant, particularly in recent years.

Research has also shown to us that, in regard to life span, it is as important to pay attention to your mind as your body. A balance of mind and body plays an important role. How important is your mind in relation to the aging process? Control of stress, relaxation, and slowing down your daily activities helps slow down aging. You must have heard people say, "Take it easy" or "Don't work too hard" and "Relax" to remind us that if we want to live longer, we must reduce anxiety and keep a relaxed mind. There is enough evidence to show that optimistic people live, on the average, 19% longer than those who are depressed and miserable in their lives.[6] You can find that an antiaging strategy requires one to stay optimistic and keep a positive outlook.

You may not realize how inactivity can affect the aging process. Other than what is normally due to "Father Time," aging can be directly related to our sedentary lifestyle. One good example of this is when an individual fractures a bone in the arm and the arm is casted. After 6–8 weeks of wearing the cast and then removing it, the muscles of the arm are thinner (atrophied) and weaker due to the inactivity. If the cast was over the knee joint, the joint would have stiffened and normal range of motion would be greatly restricted.[7] From these examples, you can perceive how important it is for us to keep our body moving. If you do not move it, you lose it. Sedentary lifestyle and inactivity affect the architectural structure and, thus, speed up the aging process.[7]

NORMAL HUMAN GROWTH AND AGING LIFE CYCLE

While getting old is an inevitable part of being alive, how fast we age is pretty much dependant on the way we live. We live in a changing world. We change every decade, every year, every month, and every day.

The human growth and aging life cycle may be broadly divided into three stages: *growth stage, peak growth stage,* and *aging stage.*[4] As shown in Figure 17.1, each of us go initially through the growth stage—infancy to childhood (0–6 years of age), youth (6–12 years of age), adolescent (12–20 years of age), and adulthood (20–35 years of age). One thing in common among these growth years is that we are living the most enjoyable years of life. Our parents, grandparents, relatives, friends, and teachers care for us as if we are in heaven or paradise. There is nothing to worry about, and we live happily days and nights, in home, or at schools.

When we reach midlife (35–45 years of age), we are at the "peak growth stage" or midlife. Living during these 10 years is totally different from the earlier growth stage. The biggest physiological change of midlife is the onset of menopause caused by weakening or even loss of production of the female or male sex hormones. Also, decreased energy levels, decreased sexual desire, decreased mental and visual acuity, decreased muscle mass, increased blood pressure, and heart disease are common. Midlife also brings

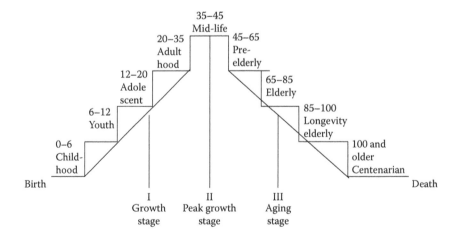

FIGURE 17.1   Normal human growth and aging life cycle.

the risk of financial crisis, because supporting a family and building up a career at the same time is a challenge. The danger of midlife is real. Midlife could be successful or a failure, and quite often is risky and dangerous. In most cases, midlife, a period after the happy enjoyable youth years, is hard work and struggle. An old American proverb says: "Mid-life is the old age of youth and the youth of old age." No man or woman is exempted from the midlife stage. It is associated with people unwilling to accept the transition of growth to aging in life and forever trying to recapture their lost youth.

The transition from youth to adulthood is easy, but from midlife to the next aging stage nothing is guaranteed.

The third, or aging stage, usually begins after midlife. From 45 to 65, this 20-year period can also be called the pre-elderly period. From 65 to 85 may be called the elderly period. From 85 to 100 is the longevity elderly period, and 100 years and above are the centenarian years.

We know that everyone experiences life differently. Aging from 65 to becoming a centenarian or to death may not be enjoyable because of physical or mental decay and debilitating changes. The increased uncertainty and hardship of the senior years can make it really miserable and cause suffering. You will find all the important information and strategies of keeping yourself in a healthy and mentally active condition in your 60s, 70s, and 80s in this book, and you probably will continue to enjoy life in the later years and become a member of the centenarian club.

WHAT IS HEALTHY AND SUCCESSFUL AGING?

Longevity has been an aspiration as long as there has been human existence on earth. According to the famous Russian physician and physiologist Eliia Michnikoff, until now human beings have not found any reliable method of prolonging life. Evidently, death is the natural end to life, and there is no being that can escape that fate. The life span differs depending on the region, climate, race, heredity, and lifestyle and has varied at different times in each historical period. The following have been considered as probable factors for successful aging and longevity.[8]

HEREDITY

The significance and influence of heredity on longevity are self-evident There are families whose members have been observed to live longer than the average human life expectancy, except in cases where death occurs because of an accident. This is because an offspring inherits strong and healthy body parts and organs from parents, including the genes of nervous, circulatory, and other systems. Modern science, however, has shown that heredity factors play only a minor role and can be neutralized by other factors such as environment, arteriosclerosis, and general health.

ENVIRONMENT

The second most important factor in longevity is having an ideal environment to live in, such as a moderate climate and clean air and water, free from harmful microbes and poisons.

Lifestyle

The type of work and working conditions as well as the duration of hours, in addition to spiritual and psychological activities have an impact on longevity. It appears that when a person enjoys good physical health and mental tranquility, it affects their life span significantly. Longevity is directly related to working habits and how one manages his or her stress under constant pressure. It is for this reason that long-term joblessness and early retirement at a younger age might actually be unhealthy and may lead to the shortening of one's life span.

Nutrition

The kinds of food and the amount we consume have an important impact on longevity. It is a fact that the majority of the people whose life span exceeded 100 years have been found to be dieters. Also, there are numerous proverbs that simplify the harm caused by overeating. Among them is this Chinese proverb: "A person digs his grave with his own teeth." In the Western world during World War I, it was observed that death as a result of diabetes had significantly declined in some countries. The main reason was that poverty as a cause of reduced intake of food is a blessing in disguise.

Health

Optimum health is designed to occur at a time of maximum productive capacity. After those productive years, the body declines, health problems become increasingly more common, and survival is threatened. You can only lengthen your life, preserve your functions, and retain a glowing attractive appearance by following an antiaging program. Look at how your friends and relatives age. Some are crippled, wrinkled, and gray haired, not as you first saw them. There are others who retain that sparkle of youth in the way they look, feel, and function into the 70s, even 80s. The difference is not due to a stroke of luck but rather is how they have taken care of their precious bodies over the years.[9]

McDougall said, "The secret for healthy, successful aging is almost too simple to believe—it's a healthy diet, moderate exercise, and clean habits. But the results are no less than a miracle."[9]

Many people are conditioned to think that after 65 years, one should sit down, relax, and watch the rest of the world with amusement and pity as it goes through the motions of detestable jobs, demanding childrearing, and dreary household chores. As Vijai P. Sharma wrote in his article, "I am immensely inspired by what Pearl S. Buck, the author of *The Good Earth* fame and Nobel Prize winner, had to say on her 80th birthday. Buck impresses me with her task-centeredness. She wrote this note sitting at her worktable facing the window and saying, 'I have much work to do and I enjoy myself and what I do.'"[10]

Pearl Buck thought about old age. She said, "I do not know what people mean when they speak of being old. I do not know because I do not know where life begins and where it ends, if there is an end. For me, death is merely the entrance into further existence."[10]

LONGEVITY TECHNIQUES IN THE ORIENT

In Early China

The pursuit of longevity in early traditional Chinese medicine (TCM) is more concerned with "nourishing life" than with reestablishing physical health that is already broken down. Even in the ancient medical text, the fundamental theme of TCM in the pursuit of longevity is in search of the perfection of the body.[11]

According to TCM, life is composed of *qi* (breath, or life energy) and blood. *Qi* is a force that manifests in respiration and that cannot be felt or seen, circulating nonstop around the body. The circulation routes of the *qi*, the conduits (or meridians as they are traditionally called in acupuncture texts) are well defined and put to use by acupuncture physicians and aspiring immortals. Blood is an essential substance for the body. It performs the physiological functions of constructing and nourishing the tissues and organs of the body. Blood circulates ceaselessly in the vessels, internally to the viscera and bowels and externally to the tissues, bones, and skin.

The Taoist encyclopedia *Yunji gigian* (*Seven Slips from a Cloudy Satchel*) contains two exemplary texts on nourishing life by means of *qi*. There are two famous books on this subject, *Fugi jingyi lun* (*Discourse on the Essential Meaning of Absorption of Qi*) and *Sheyang zhenzhong lun* (*Pillowbook of Methods for Nourishing Life*). The former was written by the ancient Taoist master, Sima Chengzhen (647–735), while the latter is attributed to Sun Simiao (581–682), the eminent Taoist alchemist and physician. Both texts say that *qi* is the main force on which to build attainment. The *qi* is the exceptional gift intrinsic to all human beings that leads them to higher accomplishments. So it is the cultivation of *qi* that allows adults to develop their willpower and proceed in the direction of immortality.[12]

Approximately at the same time, Sun Simiao also wrote *Qian Jin Yifang* (*Revised Prescriptions Worth a Thousand Ounces of Gold*) and many other medical books. All Chinese have worshiped him as the "King of Medicine." The most profound medical philosophy in relation to longevity from Sun Simiao was documented in *Shenyang Zhenzhong Lun*. In it, Sun Simiao stressed that most people think most chronic disease begins only on the day it breaks out. As a matter of fact, the origins of the disease reach back much farther and it developed gradually. He advises that the prudent practitioners who are nourishing life and longevity must therefore eliminate all diseases in their very beginning.

In addition, Sun also stresses moral aspects and mental discipline as fundamental for the quest of longevity. He emphasizes that mental effort and discipline in daily life is indispensable, otherwise all medical treatment remains ineffective and will not bring permanent betterment.[12]

The above two texts written by both Sima Chengzhen and Sun Simiao have in common that they deal with ways of avoiding illness and maintaining good health for the pursuit of longevity.

In Early Japan

Japan began to receive a strong influence of Korean medicine in about the fifth century, but Chinese medicine was transmitted directly from China to Japan only from the seventh century (Tang dynasty) onward. In the Nara period, numerous Japanese monks went to China in order to study Buddhism as well as traditional Chinese methods of healing. Famous Chinese monk, *Jianzhen* (689–763), also went to Japan to spread the dharma. Nolens volens, these monks also brought with them from China a rich knowledge of traditional Chinese[13] pharmacology, medical practices, and longevity techniques. Throughout the Nara and the first half of the Heian periods—that is well into the 10th century—Chinese medicine of the Sui and Tang dynasties occupied the position of official court medicine in Japan.

Longevity techniques are mentioned in different sections throughout *Ishinpo* (*Essential Medicine Methods*), the oldest most extensive work on Japanese traditional medicine compiled by the official acupuncturist at the imperial court, Tambe no Yasuyori (912–995). Relating to the art of nourishing life, the longevity techniques most strongly favored in Ge Hong's *Baopuzi* (*Book of the Master Who Embraces Simplicity*) deals with the fundamental principles of nourishing life. It explains that the attainment of longevity means the healing of latent disease, while perfecting the good qualities of one's character is claimed to be effective for the prevention of calamities.

The principles of nourishing life according to *Baopuzi* are as follows:

- *Fundamental principles*: Nourishing life and practicing meditation
- *Cultivating the spirit*: By nourishing the spirit one will not die
- *Nourishing the body*: Preserving one's physical body according to the four seasons
- *Applying the breath*: Be careful about environmental air one is breathing
- *Gymnastics*: Physical exercise to keep good health, energy, or breath
- *Daily regimen*: Proper deportment and activities of daily life
- *Sleep*: The correct way of sleep
- *Proper language*: Speak little, not in a loud voice
- *Clothing*: Dress properly
- *Dwellings*: Pure and simple, do not adorn or embellish them
- *Various prohibitions*: Never exhaust one's essence and *qi*; do not overeat or take too deep breaths

Following the philosophy of Tao, it is most important to nourish the spirit, and of secondary importance to nourish the physical body. The spirit should be pure and tranquil; the bone should be stable. This is the foundation of long life.

LONGEVITY AND CENTENARIANS' MYSTERIOUS LONG LIFE SECRETS

Have you ever imagined what it might be like to live to the age of 100 or over? The Chinese call living to one's maximum life *Tian nian*, meaning "living one's full life span" or maximum life span. The following text provides stories of centenarians who give their opinions as to what they did or did not do to achieve this milestone.

A remarkable article entitled, "Every day is a gift when you are over 100," appeared in the January 1973 issue of *National Geographic* magazine. It dealt with the oldest people in the world at the time, most of them over 100 years of age. In the article, Alexander Leaf, of Harvard University, found that many people living in the Caucasus of southern Russia were more than 100 years old.

If we can live healthily, it has been said that the maximum life span of the human species is about 140 years. How would you feel if you could live happily and healthily for over 100 years? Would you call it an extra gift from God? Gerontologists are sometimes asked if they really think it is a good idea for people to live this long. Obviously, the increased population of seniors would raise many socioeconomic problems with regard to their health and well-being, as well as nursing care, welfare, and disability costs. However, these problems would not be so unmanageable if the aged seniors were kept healthy. If people lived longer but were isolated, sick, disabled, and miserable, it would not be ideal.[14]

There are stories about long-lived people. Nobody can prove that they are all true. Allegedly, 20,000 centenarians live in the rural areas on the mountains of the South Soviet Union (now Russia), and some of them are even over 150 years old. Russian scientists are presently engaged in investigating the factors that lead to healthy aging and in discovering ways of prolonging life. These exceptionally long-lived people have invariably lived humble lives, doing hard physical work or exercise, often outdoors, from youth well into old age. Their diet is simple, as is their social life involving families. One example is Shisali Mislinlow who lived to be 170 years old and gardened in the Azerbaijan region in Russia. Mislinlow's life was never hurried. He said, "I am never in a hurry, so don't be in a hurry to live, this is the main idea. I have been doing physical labor for 150 years."[14]

The people of the native Abkhasians of Caucasus mountains in Soviet Russia are long lived, over 100 years, because they live a way of life that is close to nature. They use all-natural substances, no artificial or synthetic chemicals. Their foods, fruits, and vegetables are organically grown. They use herbs, such as herbal tea and herbal medicine, honey, and garlic, and do not eat much meat. They also enjoy physical activity in their advanced years.

ANTIAGING HERBS: PANAX GINSENG AND AMERICAN GINSENG

Many Chinese superior (first) grade herbs are tonic herbs; these tonic herbs, in particular, have antiaging effects. In the ancient times, all traditional Chinese medicine physicians were herbalists as well as nutritionists and acupuncturists. They taught people the *Tao* or the way to live, what to eat, and what food should be avoided for better health. Many famous physicians lived long lives. For example, the herbalist Meng Shen, who wrote China's first diet herbal, describing 227 herbs, lived to be 92 (AD 621–713). The ancient physician, Wu Pu, who wrote the *Wu Pu Ben Cao*, describing 441 herbs, lived over 100 years (AD 220–300). Sun Simiao, the most famous physician in China, who wrote the *Bei Ji Qian Jin Yao Fang* and *Qian Jin Yi Fang*, formularies together describing over 7,300 prescriptions, lived 101 years (AD 581–682). And Lan Mao, who wrote the famous *Dian Nan Ben Cao*, describing 448 herbs, many indigenous to Yunnan province, lived to be 79 years old (AD 1397–1476). Li Shi Zhen, the greatest herbalist of all time, author of the most famous Materia Medica, *Ben Cao Gang Mu*, lived 75 years (AD 1518–1593).

Chinese tonic remedies have adequate empirical proof of their effectiveness and safety. These herbs are beneficial to weakness, debility, promote everyday health, keep physiologically in balance of the body, and they are directly linked to longevity. Commonly used antiaging herbs, based on TCM theory, are tonics for energy, or *qi*, and blood, immune enhancement, antioxidants, and tonics for body's essence or *yin*, and tonics for organ function, or *yang*, of the body.

Since the early nineteenth century, attempts have been made to understand the actions and properties of Chinese medicinal herbs through scientific research. Nearly all the work has been conducted during the past 100 years, primarily in laboratories in China, Japan, Russia, and Germany. It was also during this time that most of the phytomedicines used in modern times were developed. Most of the biomedical research into the effects and uses of Chinese medicinal herbs has attempted to isolate their active ingredients and to understand their pharmacologic effects, and numerous scientific papers have been published about ginseng root (*Pananx ginseng*).[1]

As early as 2,000 years ago, the first official Chinese *Materia Medica, Shen Nong Ben Cao Jing*, states the tonic and antiaging activities of *Panax ginseng* as follows:

"Tonic to the five viscera
quieting the spirit
establishing the soul
allaying fear
expelling evil effluvia
opening up the heart and brightening the eyes
benefiting understanding
invigorating the body and prolonging life
if it is taken constantly"

In the United States, people over 65 years of age are considered aged, and people between 45 and 64 years are considered middle-aged. There are at present about 20 million Americans who are aged, and about 42 million Americans who are middle-aged. These 62 million senior citizens (constitute approximately 31% of the total population) are bound to have different physical and mental problems.[15] What are the most important factors regulating one's life when old? The National Council on Aging (NCOA) commissioned Louis Harris and Associates, Inc., to conduct a major national survey aimed at gaining a better understanding of aging and the reality of old age in the United States. The survey findings indicated that age seems to play a minor role affecting life satisfaction, but poor health is the most serious problem in the aged.[16]

Although ginseng does not confer eternal life, it is, in fact, an excellent remedy of tonification for the aging and the aged. The ancient Chinese doctors claimed: "Ginseng restores exhausted power, making old people young." It sounds unbelievable to you and me. However, more and more scientific research reveals the secrets of ginseng; ginseng indeed possesses these virtues. We know, among other effects, that (1) ginseng has antistress and adaptogenic powers; (2) ginseng stimulates the synthesis of protein, bone marrow DNA, and blood cells, thus functioning as a powerful hematinic agent; (3) ginseng is a metabolic regulator for protein, carbohydrate, and fat, including cholesterol; and (4) the tonic effect of ginseng keeps the organs (five viscera) and endocrine glands in a harmonious working order. These properties are absolutely essential in maintaining good health, thus making you feel young. We also learned that ginseng contains certain unknown principles that are capable of delaying degeneration of primary human amnion (PHA) cells *in vitro*. If the degeneration effect of cells is also delayed *in vivo*, then ginseng is indeed capable of making old people young, and we can no longer be skeptical about the Chinese herbal medicine as being based on superstition.

AMERICAN GINSENG (RADIX *PANAX QUINQUEFOLIUM*)

American ginseng is only slightly different from the plant native to Korea and China (*Panax ginseng*). American ginseng is also called *sang, red berry*, and *five fingers* by the ginseng farmers.

American ginseng is commonly used in TCM by the Chinese to replenish the Vital Essence (*Yin*), reduce the Internal Heat, promote the secretion of body fluids, for the treatment of debility, spontaneous night sweating, and chronic fatigue. American ginseng is a good antiaging medicinal herb.[17]

CHEMICAL CONTENTS

American ginseng root contains 6%–8% of total ginseng saponin, called panaquilins. Additionally, many other ingredients are similar to those found in Chinese ginseng. American ginseng root contains slightly more saponin but less proteinous and oily substances. It also contains inorganic salts, sugars, phytosterol esters, terpene, panacene, and fatty acids, as well as 18 different amino acids.

E. John Staba of the University of Minnesota and his research team in their extensive analyses of American ginseng discovered that the ginseng root contains panaquilins B, C, D, E_1, E_2, E_3, G_1, and G_2. The ingredients panaquilin G_1 and G_2, are not found in *Panax ginseng*. American ginseng contains about 17% sapongenin panaxadiol but only 0.44% panaxatriol, a ratio of about 39:1. The contents of sapongenin panaxadiol and panaxatriol in Chinese or Korean ginseng is about equal, or a ratio of 1:1. According the traditional Chinese belief, it is speculated that the higher ratio of panaxadiol to panaxatriol in the American ginseng root makes it cool, while the Chinese ginseng root is hot in nature.[14]

No doubt ginseng root is a super tonic herbal remedy as well as an antiaging herb known for thousands of years. There are many other Chinese tonic herbs that also possess antiaging activities similar to those of *Panax ginseng*. They are American ginseng, *ci wu jia*, (*wu ji ashen* or Siberian ginseng), Schisandra fruit, and astragalus root, to name just a few. Modern research and scientific studies show that these "tonic herbs" are also adaptogenic remedies defined in TCM. Adaptogenic herbs are totally different from stimulants in Western pharmacology. Adaptogens are herbs that possess a normalizing effect of the body toward various physiologic disorders. Russian scientist I. I. Brehkman defines an adaptogen as "a substance causing a state of nonspecific increased resistance (SNIR) of the organism to adverse stresses of various origin."[18]

Chinese tonic medicinal herbs selected by the author as antiaging herbs are capable of regulating and mediating the physiologic activities of the body; some of these herbs are adaptogenics in nature and are able to enhance or regulate the immune system of the body, others are herbal tonics to the *qi*, blood, yin or yang of the body, and some are powerful antioxidants.[17]

No doubt, proper use of these antiaging ginseng herbs may lead to maintenance of good health and prolonged life.

REFERENCES

1. Leaf, A., *National Geographic Magazine*, 1973, p. 93.
2. Veith, I., Ed. *The Yellow Emperor's Classic of Internal Medicine*, University of California Press, Berkeley, CA, pp. 10–18.
3. Warshofsky, F. Modern Maturity—The Methuselah Factor, November/December 1999, http://www.aarp.org?mmaturity/nov-dec99/methuselah.htm.5/15/2002
4. Zhou, J., Ed. *Study and Treatment of Gerontologial Ailments*, Huhehote, China, Innermongolia Peoples's Publishing Co., 1987, pp. 2–20.
5. Williams, R. J. *Nutrition Against Disease*, Bantam Books, New York, 1993, pp. 139–149.
6. Day, P., Rejuvenation, Longevity and Immortality, May 2002, http://www.meaningoflife.i12.com/immortality.html
7. Cane, E. M., The Aging Process, 2002.
8. al-mahdi, A.-I., The Just Leader of Humanity. The Research about Longevity—Chapter 9. https://www.al-islam.org/al-imam-al-mahdi-just-leader-humanity-ayatullah-ibrahim-amini/chapter-9-research-about-longevity
9. The McDougall Newsletter, January/February 1999. "Successful Aging," https://www.drmcdougall.com/newsletter/jan_feb.99.1.html
10. Sharma, V.P., Aging with an Attitude, 202.
11. Baptist Health System, Herbal Medicine. 2003.
12. Kohn, L., Ed., *Taoist Meditation and Longevity Techniques*, University of Michigan Press, Ann Arbor, 1989, pp. 263–296 (17).
13. Kohn, L., Ed., *Taoist Meditation and Longevity Techniques*, University of Michigan Press, Ann Arbor, 1989, pp. 1–32.
14. Hou, J. P. and Jin, Y., *The Healing Power of Chinese Herbs and Medicinal Recipes*, Haworth Press, Bingham, NY, 2005.
15. Botwinick, J., *Geriatrics*, 29: 124, 1974.
16. Beverley, E. V., *Geriatrics*, 30: 116, 1975.
17. Doug, K. et al., ed., *Contemporary Clinical Chinese Herbal Medicine*, (in Chinese) Beijing China Chinese Medicine Press, 1998, pp. 505–553.
18. Brekhman, I. I., *Man and Biologically Active Substances*, Pergamon Press, New York, 1980, pp. 58–59.

Section VII

*How to Take Ginseng and
Who Should Not Take It*

18 Take Ginseng with Care

Asian ginseng (Chinese and Korean) and American ginseng are believed to have no side or toxic effect if used correctly. Ginseng has different forms: dry root, sliced and boiled in water for a decoction with other herbs as in traditional Chinese medicine (TCM) is the commonly used form. The root after steaming and then dried becomes red ginseng.

In modern TCM, ginseng root is made in many different forms as ginseng marketable products, such as ginseng tea, ginseng drinks, capsules, powders in capsule, ginseng candies, or as an ingredient with other herbs for prescribed compounds as in TCM practice.

American ginseng can be purchased as whole roots, powder, capsules, or a liquid tincture. For whole roots, wild ginseng is the highest quality and also the most expensive. Also available are organically grown cultivated roots, which are free from pesticides and chemicals. The easiest way to prepare ginseng roots is to make a tea from them. Ginseng roots are very hard and brittle. They should be sliced and simmered in water for 45 minutes or longer to extract the majority of ginsenosides. Experts recommend avoiding metal pots, which can reduce its antioxidant properties. Some herbalists recommend boiling *ginger* or *licorice* root with American ginseng to increase its effectiveness. For each serving of tea, two or more teaspoons of ginger root are recommended.

HOW TO TAKE GINSENG

The common doses are as follows:

Asian ginseng root, usually 3–9 g for a decoction
American ginseng root, usually 3–6 g for a decoction
Siberian ginseng root, usually 9–30 g for a decoction

Precautions include the following:

- Do not give ginseng to a child.
- Women taking ginseng should consult an herbal medicine doctor first.
- Those taking Red-Asian or American ginseng should be more careful because it may cause nervousness, or sleeplessness if taken in higher doses or combined with another stimulant, such as caffeine, tea, or other drugs.
- Other side effects may occur in cases of overdose.
  - High blood pressure
  - Restlessness
  - Anxiety
  - Headache
  - Nosebleed
  - Nervousness

- Diarrhea
- Vomiting
- Skin rash
- Depression
- Sleeplessness
- Mastalgia
- Euphoria
- Breast pain

Other cautions are as follows:

- Do not take ginseng on an empty stomach to avoid hypoglycemic (low blood sugar) reactions, even in people with diabetes.
- People with high blood pressure should not take red ginseng.
- People with bipolar disease should not take ginseng because it may increase the risk of mania.
- People with autoimmune disease such as rheumatoid arthritis, lupus, or Crohn disease should take ginseng with care or consult a TCM doctor.
- Women with history of breast cancer should not take ginseng.
- Ginseng should be avoided for at least 7 days before having surgery because Asian ginseng may act as a blood thinner, which causes an increased risk of bleeding.

AMERICAN GINSENG

Pregnant women should use American ginseng only under a doctor's orders and avoid any products that contain *Panax ginseng.*

Consumers should be sure that the American ginseng product they purchase is a reputable one. Because of the high price and demand of American ginseng, some questionable products are on the market. Generally, wild American ginseng is of higher quality than the cultivated plant, and the older and larger the root, the higher the quantity of ginsenosides present. In addition, consumers should be careful not to confuse American ginseng with *Panax ginseng*, which has been shown to produce more serious adverse side effects. A 2002 report stated that the effects of *Panax ginseng* seemed more likely when the herb was used in combination with other products than when it was used alone.

Ginseng powder is also available and can be made into tea or taken with water or juice. One-half to 1 teaspoon is recommended per serving. Extracts of American ginseng are also available, in liquid or tablet form, some of which offer standardized quantities of ginsenosides. Packages of standardized products should be labeled with the appropriate dosage.

American ginseng is usually taken two to three times per day between meals. It should not be taken continuously for a long period of time.

Ginseng root can also be made into a tincture using alcohol, as ginsenosides are soluble and well-preserved in alcohol. Vodka or clear alcohol can be used, and the ginseng roots should be chopped finely or put in a blender with the alcohol. Enough alcohol should be used to completely saturate and cover the roots, and the solution should be kept in a sealed glass bottle for a month or longer. The solution should be shaken frequently to promote the extraction process. The liquid can be strained from the roots after distilling, and kept for up to 3 years. Half a teaspoon or more of the solution can be taken as a daily serving.

Side Effects

In general, American ginseng is gentle, although side effects may occur when taken in the wrong dosages, over too long a time, or by people whose constitutions, allergies, or health conditions disagree with the herb. Also, products combining Korean ginseng with American ginseng may increase the chances for side effects.[1]

REFERENCE

1. Hou, J. P. and Y. U. Jin, *The Healing Power of Chinese Medicine and Formulated Recipes*, Haworth Press, New York, 2005.

19 Possible Interaction of Ginseng with Drugs

POSSIBLE INTERACTIONS

If you are currently taking any of the following medications, you should not use Asian ginseng without first talking to a health-care provider.

ANGIOTENSIN-CONVERTING ENZYME INHIBITORS (BLOOD PRESSURE MEDICATIONS)

Asian ginseng may interact with angiotensin-converting enzyme (ACE) inhibitors used to lower high blood pressure. These medications include the following:

- Captopril (Capoten)
- Benazepril (Lotensin)
- Enalapril (Vasotec)
- Lisinopril (Prinivil, Zestril)
- Fosinopril (Monopril)
- Ramipril (Altace)
- Perindopril (Aceon)
- Quinapril (Accupril)
- Moexipril (Univasc)
- Trandolapril (Mavik)

CALCIUM CHANNEL BLOCKERS (HEART AND BLOOD PRESSURE MEDICATIONS)

Asian ginseng may make certain heart medications, including calcium channel blockers, work differently than intended. These medications include the following:

- Amlodipine (Norvasc)
- Diltiazem (Cardizem)
- Nifedipine (Procardia)

BLOOD THINNERS (ANTICOAGULANTS AND ANTIPLATELETS)

Asian ginseng may increase the risk of bleeding, especially if you already take blood thinners, such as aspirin, warfarin (Coumadin), or clopidogrel (Plavix).

CAFFEINE

Ginseng may make the effect of caffeine stronger, possibly causing nervousness, sweating, insomnia, or irregular heartbeat.

DIABETES MEDICATIONS, INCLUDING INSULIN

Ginseng may lower blood sugar levels, increasing the risk of hypoglycemia or low blood sugar.

DRUGS THAT SUPPRESS THE IMMUNE SYSTEM

Asian ginseng may boost the immune system and may interact with drugs taken to treat an autoimmune disease or drugs taken after organ transplant.

STIMULANTS

Ginseng may increase the stimulant effect and side effects of some medications taken for attention deficit hyperactivity disorder (ADHD), including amphetamine and dextroamphetamine (Adderall) and methylphenidate (Concerta, Ritalin).

MONOAMINE OXIDASE INHIBITORS

Ginseng may increase the risk of mania when taken with monoamine oxidase inhibitors (MAOIs), a kind of antidepressant. There have been reports of interaction between ginseng and phenelzine (Nardil) causing headaches, tremors, and mania. MAOIs include the following:

- Isocarboxazid (Marplan)
- Phenelzine (Nardil)
- Tranylcypromine (Parnate)

MORPHINE

Asian ginseng may block the painkilling effects of morphine.

FUROSEMIDE (LASIX)

Some researchers think Asian ginseng may interfere with Lasix, a diuretic (water pill) that helps the body get rid of excess fluid.

OTHER MEDICATIONS

Asian ginseng may interact with medications that are broken down by the liver. To be safe, if you take any medications, ask your doctor before taking Asian ginseng.

VERATRUM NIGRUM

The herb cannot be used simultaneously with ginseng.[1-3]

REFERENCES

1. Adam, L. L. and Gatchel, F. J. "Complementary and Alternative Medicine: Application and Implication for Cognitive Functioning in Elderly Population," *Alt. Ther.*, 7(2): 52–69, 2000.
2. Ang-Lee, M. K., Moss, J., and Yuan, C. S., "Herbal Medicine and Perioperative Care," *JAMA*, 2: 208–216, 2001.
3. http://www.umm.edu/health/medical/altmed/herb/asian-ginseng

Glossary

Acupuncture: The ancient practice, especially as carried on by the Chinese, of piercing parts of the body with needles in seeking to treat disease or relieve pain.

Adaptogen: A substance causing a state of nonspecific increased resistance (SNIR) of the organism to adverse stresses of various origins.

Adenoma: A benign tumor of glandular origin.

Adrenalectomy: Surgical removal of an adrenal gland.

Aglycone: The nonsugar portion of a glycoside.

Alkaloids: Nitrogenous crystalline or oily compounds, usually basic in character, such as atropine, morphine, quinine, etc.

Alternative (as a drug): Gradually changing, or tending to change, a morbid state of the functions to a healthy person.

Amnion cells: Cells composing the thin, translucent wall of the fluid-filled sac for the protection of the embryo.

Analog: Part having the same function as another but differing in structure and origin.

Androgen: A male sex hormone or synthetic substance that can give rise to masculine characteristics.

Anemia: A condition in which there is a reduction of the number of red blood corpuscles or of the total amount of hemoglobin in the bloodstream or both, resulting in paleness, generalized weakness, etc.

Antagonism: A mutually opposing action that can take place between organisms, muscles, drugs, etc.

Antagonistic: Showing antagonism; acting in opposition.

Anxiety: A state of being uneasy, apprehensive, or worried about what may happen; in psychiatry, an intense state of this kind, characterized by varying degrees of emotional disturbance and psychic tension.

Aphrodisiac: Any drug or other agent arousing or increasing sexual desire.

Apoplexy: Sudden paralysis with total or partial loss of consciousness and sensation, caused by the breaking or obstruction of a blood vessel in the brain; stroke.

Arthritis: Painful inflammation of a joint or joints of the body, usually producing heat and redness. The condition can be brought about by nerve impairment, increased or decreased function of the endocrine glands, or degeneration due to age.

Ataractic: A tranquilizing drug; of or having to do with tranquilizing drugs or their effects.

Atherosclerosis: A thickening and loss of the elasticity in the inner walls of the arteries and accompanied by the deposition of atheromas or fatty nodules.

Biological assay: A means of estimating the strength or potency of a drug by using some living organism or animal. The strength of the unknown sample is compared with a known or standard drug.

Callus: A mass of undifferentiated cells that develops over cuts or wounds on plants as at the ends of stem or leaf cuttings.

Cancer: Common term for a neoplasm, or a tumor, that is malignant. A large portion of human cancers may be caused by various chemicals, such as nitrites, some steroids, asbestos, smoking, radiation, viruses, etc.

Carcinogen: Any agent or substance that produces or causes cancer.

Carcinoma: Any of several kinds of cancerous growths made up of epithelial cells.

Cardiovascular: Referring to the heart and the blood vessels as a unified body system.

Carminative: Having the power to relieve flatulence and colic.

Castration: The removal of the testicles; gelding.

Cerebrovascular: Referring to the brain and the blood vessels as a unified body system.

Chemotherapy: The prevention or treatment of infection by the systemic administration of chemical drugs.

Cholesterol: A sterol, or fatty alcohol, found especially in animal fats, blood, nerve tissue, and bile and thought to be a factor in atherosclerosis.

Choline: A viscous liquid ptomaine, found in many animal and vegetable tissues: a vitamin of the B complex.

Chromatography: A method of analysis in which the flow of the solvent or gas promotes the separation of substances by differential migration from a narrow initial zone in a porous sorptive medium. Four types generally employed are column, paper, thin-layer, and gas.

Component: An element or ingredient; any of the main constituent parts.

Constituent: An essential part, component. The chemical entities contained in a crude drug.

Consumption: A disease that causes the body or part of the body to waste away; especially tuberculosis of the lungs.

Convulsion: A violent, involuntary contraction or spasm of the muscles.

Crude drugs: Naturally occurring materials of animal, plant, and mineral origin that have not been chemically processed or purified.

Debilitating disease: Any disease that causes a weakening of a patient.

Debility: Weakness, feebleness, languor of body.

Degeneration: Deterioration of mentality; deterioration in structure or function of cells, tissues, or organs, as in disease or aging.

Demulcent: Medicine or ointment that soothes irritated or inflamed mucous membrane.

Depressant: A drug or medicine that lowers the rate of muscular, nervous activity.

Depression: An emotional state of dejection usually associated with manic-depressive psychosis.

Depressor: A nerve, stimulation of which by an agent, lowers arterial blood pressure by reflex vasodilation and by slowing the heart.

Diabetes: An inheritable, constitutional disease of unknown cause, characterized by the failure of the body tissues to oxidize carbohydrate at a normal rate. Its most important factor is a deficiency of insulin.

Diastase: An enzyme from malt that converts starch to maltose by hydrolysis.

Dispensatory: A book containing a systematic discussion of medicinal agents, including origin, preparation, description, uses, and modes of action.

DNA (desoxyribonucleic acid): A type of nucleic acid, found in animal and plant cells, occurring in the nuclei; it contains phosphoric acid, D-2-desoxyribose, adenine, guanine, cytosine, and thymine. The substance from normal cells appears to differ from that of cancer cells.

Dosage form: The physical state in which a drug or drugs is dispensed, such as tablets, capsules, or injectables, suitable for drug delivery to the patient.

Dose: The amount of drug needed at a given time to produce a particular or clinically desired activity or effect.

Dyspepsia: Disturbed digestion, indigestion, impaired digestion.

Edema: Excessive accumulation of fluid in the tissue spaces; due to disturbance in the mechanisms of fluid exchange.

EEG (electroencephalogram): A tracing showing the changes in electric potential produced by the brain.

Eleutheroside: A biologically active saponin glycoside, isolated from the root of Siberian ginseng, having varied but similar, activities of ginseng glycosides.

Endocrine glands: Any of the ductless glands, such as the adrenals, the thyroid, the pituitary, whose secretions pass directly into the bloodstream.

Enzyme: A catalyst of protein nature, which accelerates biological reactions but remains apparently unchanged itself, when the reaction is completed.

Erythrocyte: A red blood corpuscle; it is a very small, circular disk with both faces concave, and contains hemoglobin, which carries oxygen to the body tissues.

Estrogen: Any of a group of female hormones. The estrogens cause the thickening of the lining of the uterus and vagina in the early phase of menstruation; responsible for female secondary sex characteristics. Estradiol, estrone, and estriol account for most of the estrogenic activity.

Excipient: An inert substance or substances used to give a pharmaceutical dosage form suitable for delivery.

Expectorant: A remedy that promotes or modifies the amount of fluid or semifluid matter from the lungs and air passages expelled by coughing and spitting.

Fatigue: Inability to perform reasonable and necessary physical and/or mental activity. Fatigue may be associated with systemic disorders such as anemia, deficiency of nutrition, oxygen, addiction to drugs, endocrine gland disorders, or kidney disorders in which there is a large accumulation of waste products, or psychic disorders, etc. Excess fatigue causes exhaustion.

Flavonoid: Any of the flavone derivatives, including citrin, hesperetin, hesperidin, rutin, quercetin, and quercitrin, which may reduce capillary fragility in certain cases.

Gastroenteritis: Inflammation of stomach and intestines.

Genin: The aglycone or nonsugar portion of glycosides in plants.

Ginsengenin: The genin or aglycone of *Panax ginseng* extracted by alcohol.

Ginsenoside: Japanese term referring to a number of ginseng saponin glycosides isolated from the methanol extract of *Panax ginseng* root.

Glycoside: Substance that on hydrolysis yields one or more sugars and genin, or aglycone. The sugar portion is called glycone. The most important glycosides-containing plants are digitalis, rhubarb, aloe, glycyrrhiza, ginseng, etc.

Glycosuria: The presence of sugar in the urine.

Gonadotropic: A substance (hormone) that is gonad-stimulating.

Hematinic: A blood tonic that increases the formation of hemoglobin and red blood cells.

Hemodynamics: The study of how the physical properties of the blood and its circulation through the vessels affect blood flow and pressure.

Hemoglobin: The red coloring matter of the red blood corpuscles, a protein yielding heme and globin on hydrolysis; it carries oxygen from the lungs to tissues and carbon dioxide from the tissues to the lungs.

Hemolytic: Referring to the destruction of red blood cells and the escape of hemoglobin.

Herbal: Referring to a plant used for medicinal purposes or for its odor or flavor.

Histamine: An amine produced by the decarboxylation of histidine and found in all organic matter; it is released by the tissues in allergic reaction, lowers the blood pressure by dilating blood vessels, stimulates gastric secretion, etc.

Histaminic: Causing the stimulation of the visceral muscles, dilation of the capillaries, stimulation of the salivary, pancreatic, and gastric secretions.

Homeostasis: The maintenance of steady states in the organism by coordinated physiologic processes. Thus all organ systems are integrated by automatic adjustments to keep within narrow limits disturbances excited by the changes in the organism or in the surroundings of the organism.

Hormone: Active principles secreted by endocrine glands: epinephrine, thyroxin, insulin, estradiol, testosterone, etc.

Hyperglycemia: Excess of sugar in the blood.

Hyperglycemic: Referring to the condition, hyperglycemia.

Hypertension: Excessive tension, usually synonymous with high blood pressure.

Hypertensive: Referring to hypertension.

Hypnotic: A remedy that causes sleep; inducing sleep.

Hypoglycemia: The condition produced by a low level of sugar in the blood.

Hypoglycemic: Referring to the condition hypoglycemia.

Hypotension: Diminished or abnormally low tension, usually synonymous with low blood pressure.

Hypotensive: Referring to hypotension.

Impotence: Inability of the male to perform sexual intercourse. Impotence may result from physical causes such as structural abnormalities of the genital organs; decreased activity of the thyroid, pituitary, or sex glands; anemia or other debilitating diseases; alcoholism; or may be psychological in origin.

Inflammation: The reaction of the tissues to injury. The essential process, regardless of the causative agent, is characterized clinically by local heat, swelling, redness, and pain.

Insomnia: Sleeplessness; disturbed sleep; a prolonged condition of inability to sleep.

Intraperitoneal: Within the peritoneum or peritoneal cavity.

In vitro: In glass; referring to a process or reaction carried out in a culture dish, test tube, etc.

In vivo: In the living organism: used in contrast to *in vitro*.

Ketoacidosis: Acidosis accompanied by an increase in the blood of such ketones as β-hydroxybutyric, and acetoacetic acids.

LD$_{50}$ (lethal dose 50): The dosage by which 50% of the experimental animals die.

Leukemia: Any disease of the blood-forming organs, resulting in an abnormal increase in the production of leukocytes often accompanied by anemia and enlargement of the lymph nodes, spleen, and liver.

Leukocyte: One of the colorless, more or less ameboid cells of the blood, having a nucleus and cytoplasm.

Leukocytosis: Increase in the leukocyte count above the upper limits of normal.

Leukopenia: A decrease below the normal number of leukocytes in the peripheral blood.

Longevity: The length or duration of the life; used to indicate an unusually long life.

Malaise: A general feeling of illness, lack of appetite, and decreased energy.

Mast cells: A cell containing large, easily dye-stained granules found in connective and other body tissues.

Materia Medica: The division of pharmacology that treats the sources, descriptive, and preparations of substances used in medicine.

Metabolism: Sum of all biochemical processes involved in life, two subcategories of metabolism are anabolism, the building up of complex organic molecules from simpler precursors, and catabolism, the breakdown of complex substances into simpler molecules, often accompanied by the release of energy; metabolic reactions are usually catalyzed by enzymes.

Muscarinic: Stimulation of the parasympathetic nerves and slowing of the heartbeat; increasing the secretions of the salivary.

Nephritis: Inflammation of the kidney.

Neurasthenia: A group of symptoms ascribed to debility or exhaustion of the nerve centers; fatigability, lack of energy, various aches and pains, and disinclination to activity.

Neurogenic: Of nervous origin; stimulated by the nervous system.

Neurosis: A disorder of the psyche of psychic functions.

Normalization: Reduction to normal or standard state.

Oncology: The study or science of neoplastic growth (cancer).

Organism: Any living entity having differentiated members with specialized functions that are interdependent and that is so constituted as to form a unified whole capable of carrying on life processes.

Panaquilin: Group of nine ginseng saponin glycosides successfully isolated and identified from American ginseng root.

Panaxadiol: A substance (genin) yielded from panaxosides after acid hydrolysis.

Panaxatriol: A substance (genin) yielded from panaxosides after acid hydrolysis.

Panaxoside: Soviet term referring to a number of ginseng saponin glycosides isolated from the methanol extracts of *Panax ginseng* roots.

Papavarine-like: Having a local anesthetic or muscle-relaxing effect.

Paralysis: Partial or complete loss or temporary interruption of a function, especially of voluntary motion or of sensation in some parts or all of the body.

Pen-ts'ao: The compendium dealing with crude drugs. The first *Pen-ts'ao* is called *Shen-nung Pen-ts'ao ching,* formally published in the second century. The most complete *Pen-ts'ao* is *Pen-ts'ao Kang-mu*, published by Li Shih-chen in 1596.

pH: A chemical symbol used to express acidity and alkalinity in terms of the hydrogen ion concentration. The pH values may range from 0 to 14; numbers less than 7 indicating acidic, and numbers greater than 7 indicating basic.

Pharmacognosy: The science dealing with the preparation, uses, and properties of crude drugs. In a broad sense, pharmacognosy embraces the knowledge of the history, distribution, cultivation, collection, selection, preparation, commerce, identification, evaluation, preservation, and use of drugs and other agents affecting the health of man and animals.

Phospholipid: A type of lipid compound that is an ester of phosphoric acid and contains, in addition, one or two molecules of fatty acid, an alcohol, and a nitrogenous base, such as lecithin, cephalin, and sphingomyelin.

Phytosterol: Any of several steroid alcohols found in plants.

Placebo: A biologically inert substance, such as lactose, that is used as a sham drug. The placebo has no inherent pharmacological activity but may produce a biologic response.

Polyuria: Excessive urination, as in some diseases.

Pressor: Designating a nerve that, when stimulated, causes a rise in blood pressure; a substance capable of raising blood pressure.

Psychasthenia: A group of neuroses characterized by phobias, obsessions, undue anxiety, etc.

Psychopharmacology: The study of the actions of drugs on the mind.

Pulse: Alternate expansion and contraction of artery walls as heart action varies blood volume within the arteries. Usually, the pulse rate is determined by counting the pulsations per minute in the radial artery at the wrist. Various diseases may be indicated by changes in the rate, rhythm, and force of the pulse.

Pulsology: The study and science of pulse.

Pyretic: Of, causing, or characterized by fever.

Rejuvenation: A renewal of youth; a renewal of strength and vigor; specifically, a restoration of sexual vigor.

Rf value: The ratio of the distance traveled on the plate or paper by the test substances to the distance traveled by the solvent front of the mobile phase, from the point of application of the test substance.

Rheumatism: Any of various painful conditions of the joints and muscles, characterized by inflammation, stiffness, etc., and including rheumatoid arthritis, bursitis, neuritis, etc.

Rheumatoid arthritis: A chronic disease whose cause is unknown, characterized by inflammation, pain, and swelling of the joints accompanied by spasms in adjacent muscles and often leading to deformity in the joints.

RNA (ribonucleic acid): Nucleic acid occurring in cell cytoplasm and the nucleolus, first isolated from plants, but later found also in animal cells, containing phosphoric acid, D-ribose, adenine, guanine, cytosine, and uracil.

Sapogenin: The nonsugar portion of ginseng saponin glycosides.

Saponin: Characterized by forming colloidal solutions in water that foam upon shaking, have a bitter, acrid taste, and are irritating to the mucous membrane; hemolytic to blood cells.

Sedative: Quieting function or activity.

Shock: Clinical manifestations of defective venous return to the heart with consequent reduction in cardiac output. It may be caused by inadequate pumping by the heart, by reduction of the blood volume due to dehydration, or to loss of blood or plasma, or by reduced blood pressure resulting from dilation of the blood vessels.

Side effect: Drug-induced symptom that may be undesirable.

Stigmasterol: A sterol derived from the soybean.

Stimulant: An agent that causes increased functional activity.

Stomachic: One class of substances that may stimulate the secretory activity of the stomach.

Stress: Any stimulus or succession of stimuli of such magnitude as to tend to disrupt the homeostasis of the organism.

Synergism: When the response or action of one drug is enhanced by the other, synergism occurs. The term *potentiation* has been used for synergism.

Synergistic: Referring to synergism.

Thin-layer chromatography (TLC): Characterized by the application of dry, finely powdered absorbent in a thin, uniform layer to a glass plate. This plate is comparable to an open chromatographic column, and the separation of the components in the test material is based on absorption, partition, depending on the absorbents and the solvent used. Identification of an unknown component is made by comparing its Rf value with that of the known (standard) material.

Tonic: An agent or drug given to improve the normal tone of an organ, or of the patient generally.

Tonic effect: Mentally or morally invigorating; stimulating.

Trypanosoma: Slender, elongate organisms with a central nucleus, posterior blepharoplast, and an undulatory membrane, from which a free flagellum projects forward.

Vitality: Mental or physical vigor; energy.

Volatile oils: Essential oils that represent the odoriferous principles of plants: peppermint oil, clove oil, rose oil, etc.

Index